EARTH

THE POWER OF THE PLANET

IAIN STEWART & JOHN LYNCH

EARTH

THE POWER OF THE PLANET

BBC
BOOKS

Contents

'Suddenly, from behind the rim of the moon, in long, slow-motion moments of immense majesty, there emerges a sparkling blue and white jewel, a light, delicate sky-blue sphere laced with slowly swirling veils of white, rising gradually like a small pearl in a thick sea of black mystery. It takes more than a moment to fully realize this is Earth... home.'

US ASTRONAUT EDGAR MITCHELL

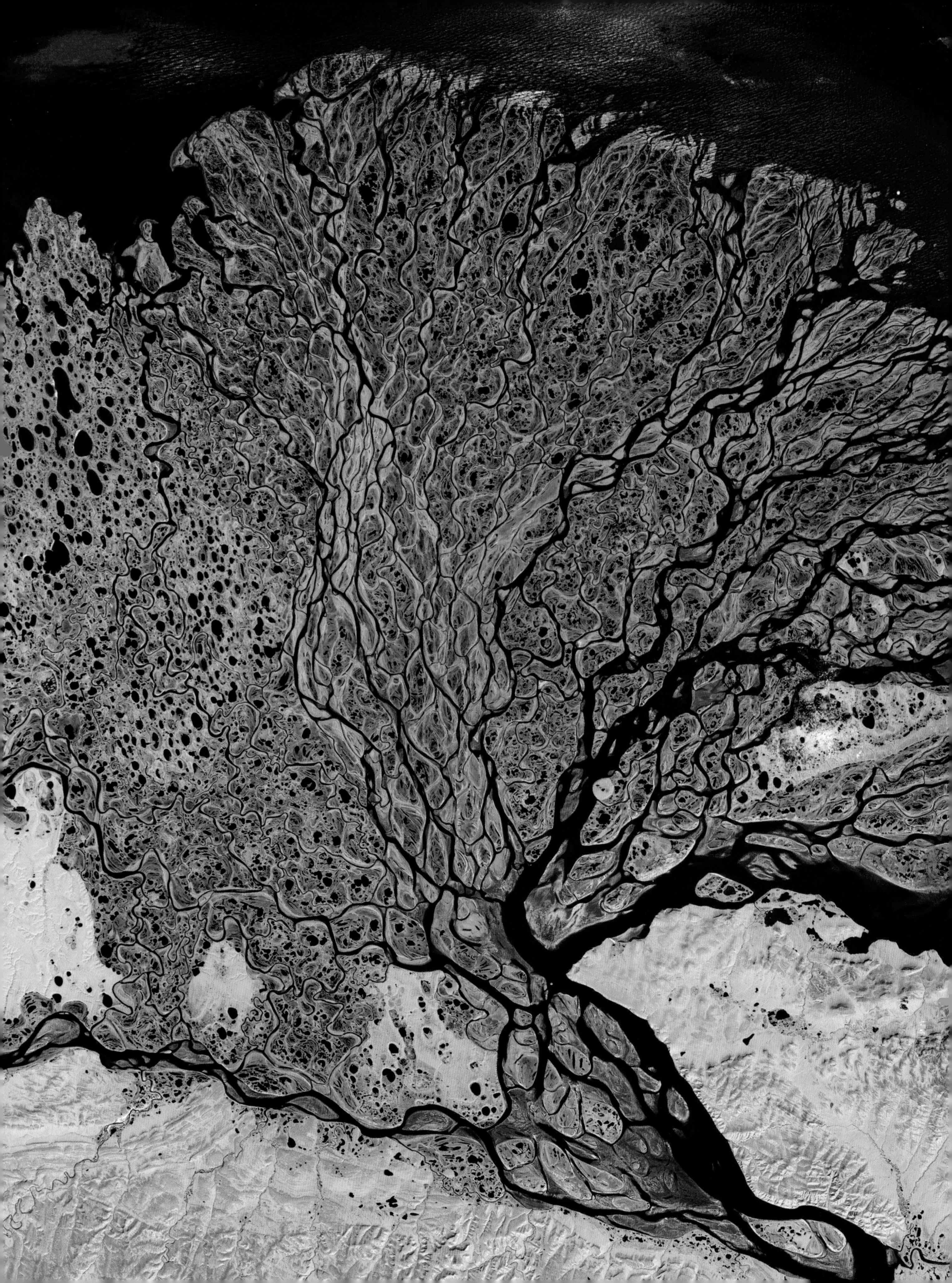

Prologue

Cherskii, northeastern Russia. A frigid outpost in one of the bleakest parts of the planet: the vast, featureless expanse of mud, moss and marsh that is the Siberian tundra. The town sits on a peat bog that is frozen for most of the year and as large as France – yet inhabited by only 20,000 people, many of them indigenous Yakuts who scratch out a living by herding reindeer. Few places on Earth, outside Antarctica, are emptier. So bitterly cold and remote is the region that Stalin (and, before him, the rulers of Imperialist Russia) chose it for their notorious 'gulag' prison camps. Little effort was needed to confine the prisoners – there was nowhere to run to. Today, the prison camps have gone but the land remains punishing. You need a very good reason to come here, and for scientists there is one, for this land may hold the key to the future of our planet.

During the brief Siberian summer, the frozen ground thaws and the tundra turns into a waterlogged patchwork of wetlands and lakes. On the outskirts of Cherskii is Blood Lake, its name a grim legacy of the area's gulag past. In winter, like all the lakes here, Blood Lake freezes over, its waters concealed by a metre or so of opaque, greyish-white ice that can safely take the weight of a person or a snowmobile. Visit in late spring, however, and the frozen crust is thinner and

OPPOSITE THE VAST LENA RIVER DELTA IN SIBERIA – A PATCHWORK OF BRAIDED RIVER CHANNELS POCKMARKED WITH THAW LAKES. THIS CORNER OF NORTH-EASTERN SIBERIA MAY HOLD THE KEY TO THE FUTURE OF OUR PLANET.

weaker, with dark patches around the lake margin where the thinning ice has become glassy and transparent. Large pockets of air can be seen trapped inside the ice here, and streams of tiny bubbles can be seen percolating up from the depths – almost as if Blood Lake is in suspended animation, breathing faintly as it waits to be released from its wintry tomb. But these are no ordinary air bubbles. If you pierce the ice and hold a flame to the gas that hisses out, an incandescent jet shoots up. For a minute or two, flames as high as the surrounding larch trees illuminate the polar night as several hundred litres of gas burn off. Step a few metres away, and you soon find other pockets of gas where the experiment can be repeated. In fact, virtually all the lakes in the region emit the same strange vapour: an invisible, odourless and incredibly flammable gas. The Siberian tundra is leaking methane.

Methane is one of the planet's most potent greenhouse gases. It might not be as famous as carbon dioxide, but it is far more virulent: weight for weight, it traps 21 times as much heat in the atmosphere as carbon dioxide does, which means that even relatively small releases of methane could have a drastic impact on climate. Since the last ice age, Siberia's methane has been trapped within the deep layer of frozen ground, or 'permafrost', that underpins the entire tundra region. Only the uppermost half metre or so of soil thaws here in summer and refreezes in winter, while the earth beneath stays permanently frozen, in places to more than a kilometre deep. In the 10,000 years since the ice age, very little methane has escaped from this frigid land, but now it seems that the heat has been turned up.

Temperatures are rising faster in Arctic lands than anywhere else on the planet. In Alaska, spring now arrives two weeks earlier than it did half a century ago, with air temperatures in the interior having risen by 2°C (3.6°F) since 1950 and permafrost temperatures up 2.5°C (4.5°F). Siberia may be warming even faster, its average air temperature up perhaps 3°C (5.4°F) in the last decade, and here too spring has sprung roughly a day earlier each year since satellite measurements began in the early 1980s. Across Alaska, northern Canada and Siberia, permafrost is turning to mush, causing roads and buildings to subside and fracture as their formerly solid foundations give way. Frozen rivers, which for most of the year serve as roads, can be driven on for ever briefer periods. More and more land is becoming unmanageable bog for longer and longer each year. And as Siberia's permafrost thaws, its methane is spewing out.

The Siberian permafrost is estimated to hold a quarter of all of the methane stored in the world's land, and the thawing lakes are letting that reservoir seep out at an increasing rate. To make matters worse, as snow cover declines across the warming tundra, the thaw is likely to accelerate, since bare soil tends to absorb the Sun's warmth rather than reflecting it back into space, as snow does. A giant burp of methane could be on the cards, and when it goes, there will be no stopping it – there are no brakes we can apply.

It sounds like just another global-warming horror story. Except that many geologists believe that something similar has happened before. Around 55 million years ago – long before the ice ages – trillions of tonnes of methane are thought to have escaped from sea-floor sediments and built up in the atmosphere over a few thousand years, causing the planet to warm up dramatically. The average air temperature at Earth's surface rose from 18°C (64°F) to 24°C (75°F); for comparison, the current global temperature is about 15°C (59°F). The entire world was transformed. Lakes evaporated, seas vanished. Forest turned to scrubland, scrubland turned to desert. The poles melted, and the Arctic Ocean became as warm as today's tropical seas. Here inside the Arctic Circle, where little more than lichen

OPPOSITE **THE FROZEN GROUND OF NORTH-EASTERN SIBERIA IS MELTING, AND FROM ITS COUNTLESS THAWING LAKES – SUCH AS HERE BETWEEN THE MAYN AND ANADYR RIVERS – DANGEROUS METHANE IS LEAKING OUT.**

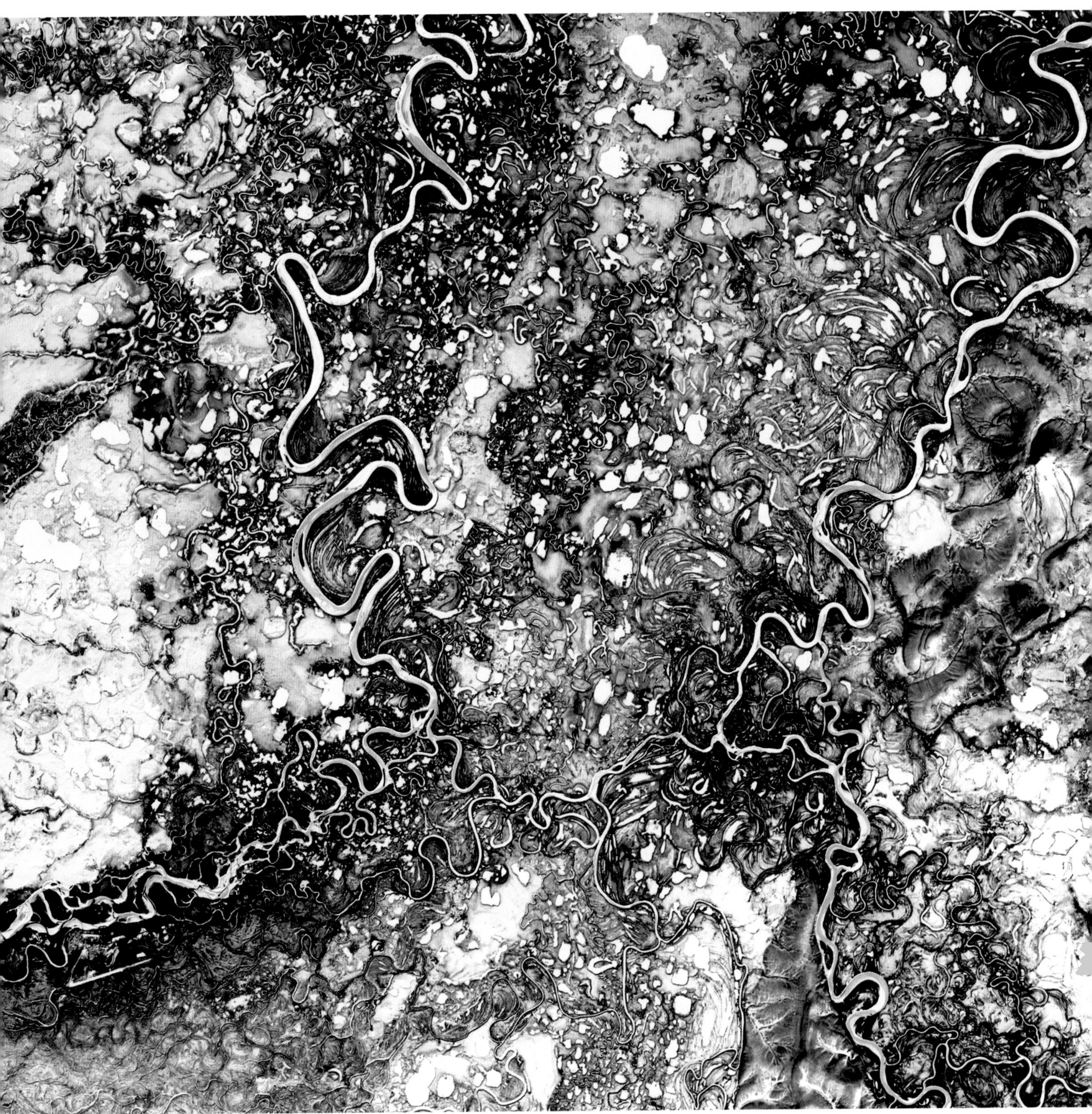

and moss can survive today, are the remains of pine forests that blossomed and flourished in the wave of warmth. Animals migrated into new continents, reaching areas that had been cut off by ice or water. But there was a downside too. The oceans were poisoned by the change in climate, and their depths were starved of oxygen. Marine organisms died by the million, and many species became extinct.

Today the greenhouse gases we are pumping into the atmosphere are pushing up temperatures even faster than they rose 55 million years ago. If the changing climate also unlocks stores of methane trapped in Arctic permafrost or deep-sea sediments, then global warming could accelerate beyond all expectations. The message from the past is clear: it has happened before, and it could happen again.

It is only in the last few decades that scientists have begun to discover just how sensitive our planet may be to the changes we are forcing upon it. The equable climate that makes life on Earth possible depends on a hidden web of dynamic connections between ocean, land and atmosphere. These ostensibly separate parts of our world in fact work intimately together like cogs in a complex machine, contriving through an endless exchange of raw materials and energy to keep the climate in balance and the planet hospitable. The problem is that our interconnected world is not only complex but also fickle. An apparently small change to one part of the system – such as methane leaking from thawing Siberian bogs – can, through a chain-reaction of interlinked processes, have profound global consequences.

A Revolution in the Air

In the 1760s, the Scottish instrument maker James Watt began tinkering with an ingenious but rather ineffective 'atmospheric engine', unaware that he was on the brink of ushering in not simply a new industrial age but a new geological one. Watt's improved design – the 'steam engine' – would set crackling a chain reaction that led to an explosion in the use of coal and iron, the mass production of machined goods and the invention of revolutionary new forms of transport. But even as the technological know-how of the Industrial Revolution spread at breakneck speed around the globe, its more insidious effects were beginning to accumulate. In the icy wilderness of Antarctica and Greenland, layers of snow dating from the 1780s record the start of a gradual but clear rise in the greenhouse gases carbon dioxide and methane. Around the same time, freshwater life in our lakes began to adjust to the new global chemistry. Many scientists now identify the latter quarter of the eighteenth century as the point when the human impact on the environment became global. Having lived at the mercy of Earth's natural forces for millennia, humanity was now beginning to exert its dominance over the planet. We had entered what some scientists define as a new geological era: the Anthropocene – the Age of Humankind.

Even in the white heat of industrial expansion, some foresaw how our world might be about to change. In fact, the greenhouse gas idea, which seems such a modern concept, was first put forward in 1827 by a French mathematician and physicist, Joseph Fourier. Fourier recognized that there is an imbalance in the amount of heat that Earth absorbs from the Sun and the amount it reflects back into space. He realized that part of this imbalance is caused by the atmosphere, which traps some of the reflected heat and so makes the planet's surface warmer. We now know that this greenhouse effect is vital for life – without it, the global temperature would be –18°C (0°F) rather than 15°C (59°F). Fourier was the first to speculate that human activity might influence how much heat the atmosphere retained, though he could not quite see how. The English scientist John Tyndall got closer to the answer in the 1850s when he worked out that much of the heat is absorbed by water vapour and carbon

OPPOSITE **GLOBAL CLIMATE CHANGE SIMULATIONS PREDICT STORMY TIMES AHEAD.**

dioxide – gases that together make up a mere 1 per cent of the atmosphere.

Today, we have a good idea of what is happening to carbon dioxide levels in the atmosphere because they have been precisely recorded for almost 50 years at an observatory on the summit of Mauna Loa volcano in Hawaii. Since 1958, this window on the sky has gathered compelling evidence that our planet is changing. Carbon dioxide levels have risen steadily by some 22 per cent over the last half century, and they are still climbing by almost 2 parts per million (ppm) each year, reaching 384 ppm in February 2007. In other words, there are currently 384 molecules of carbon dioxide in every million molecules of dry air. It might seem a piffling amount, but it represents a 37 per cent rise since the pre-industrial era, when the carbon dioxide level was a mere 280 ppm. We know that from measuring the chemicals inside ancient bubbles of air trapped in Antarctic and Greenland ice. In fact, this frozen archive of the atmosphere goes back at least 650,000 years, and as far back as we can reliably measure, carbon dioxide levels have never been higher than today.

It isn't just carbon dioxide that is changing Earth's climate. Since Arrhenius happily predicted his better,

BELOW **FARMING STYLES MAY VARY GREATLY AROUND THE WORLD, BUT THE SCALE OF AGRICULTURE'S IMPACT ON THE LAND IS EQUALLY DRAMATIC ON EVERY CONTINENT.**

The Carbon Calculator

It was a portly, party-loving Swede, Svante Arrhenius, who in 1896 made the great leap for climate science by calculating the precise effect that carbon dioxide could have on climate. Weighing almost 100 kg (220 pounds), Arrhenius was a huge man with a huge idea: he believed that ice ages were caused by vast fluctuations in the level of carbon dioxide in the atmosphere. Fixed on this idea, the Nobel Prize winner set about calculating how much carbon dioxide was needed to change the global temperature. It was a Herculean task; even today's most powerful computers can take a week or more to model changes in global climate. Arrhenius had to do the fiendishly complex calculations by hand. He worked out the warming effect of five different levels of carbon dioxide at every 10 degrees of latitude north and south on the globe. Working obsessively, often 14 hours a day, he spent over a year on the task and arrived at a surprisingly accurate set of tables: a kind of ready-reckoner for climate change. He predicted that a doubling of the carbon dioxide level would produce a 5°C (9°F) increase in global temperature, which turns out to be pretty close to the figure that most scientists currently accept. That might not seem dramatic, but, to put it in context, it's the difference in temperature between an ice age and a warm period like today.

The great contribution of Svante Arrhenius was to make the link between the burning of fossil fuels and a warming effect on the climate. Again, he laboriously calculated how long it would take for the burning of coal to double the carbon dioxide level, and, based on the industrial machinations of the late 1890s, his answer came out at about 3000 years. Ironically, Arrhenius did not see global warming as a threat; instead he foresaw an era when our distant descendants would bask in a warmer and less harsh climate than that of his compatriots in Scandinavia. What he did not foresee was that the output of carbon dioxide would increase fifteenfold over the next century. Today, society coughs and wheezes out 6000 million tonnes (6 'gigatonnes') of carbon dioxide every year worldwide. At that rate, the carbon dioxide level will reach twice the pre-industrial value towards the end of the twenty-first century. Sadly, the timing of Arrhenius's prediction was out by a factor of ten.

warmer world, the arrival of the motor car, new industries and intensive agriculture practised on a scale that was unimaginable a century ago have all led to the release of even more potent greenhouse gases, such as nitrous oxide and methane. The result, all in all, is that there has been an increase in the mean atmospheric temperature near the planet's surface of 0.7°C (1.3°F) since the start of the twentieth century.

Global Health Check

In the same way that a healthy human body needs its heart, brain, lungs and so on to function properly, so the organs of the Earth – its land, its oceans, its atmosphere, its ice (cryosphere) and its life (biosphere) – need also to be in good working order. To find out if they are, the vital signs of the planet are routinely monitored by a global network of satellites, aircraft and ground-based instruments that record things like air temperature, evaporation from the oceans, melting and formation of ice, and carbon release from forests. And so it is that every few years, the resulting mass of data and findings are fed into a planetary health check carried out by the Intergovernmental Panel on Climate Change (IPCC). The latest one, published in 2007, makes gloomy reading.

According to the IPCC, global warming is now 'unequivocal' and human activity is 'very likely' to blame. Warming of ocean waters has reached depths of at least 3000 metres (9800 feet), making sea water expand in the process and causing ocean levels to rise. Mountain glaciers are retreating and snow cover is waning in both hemispheres, and the consequent meltwaters are also adding to rising seas. Throughout Earth's vital functions, change is detectable. Rainfall patterns, wind intensities and temperature trends are deviating from established historical norms in different parts of the globe. The first signs of biological and

LEFT **THE AMAZON RAINFOREST HAS SURVIVED FOR TENS OF MILLIONS OF YEARS BUT IT COULD BE IRREPARABLY DAMAGED WITHIN THE NEXT CENTURY.**

Prognosis for a Planet

So what is the prognosis for the planet? Well, on current trends, average global temperatures can be expected to rise by 2–3°C (3.6–5.4°F) within the next 50 years or so. That will bring Earth to a temperature that is not just much higher than anything previously witnessed by human civilization but higher than anything the planet has experienced for 3 million years. By the end of the century, temperatures are likely to be up by 4.5°C (8.1°F) or perhaps more. What this amount of warming will mean for the world of AD 2100 isn't entirely clear, but global climate models – the crystal balls of climate science – provide some probable scenarios. Earth will probably have an ice-free Arctic, since high latitudes will bear the brunt of the rising temperatures. For a global average warming of 4°C (7.2°F), equatorial oceans and coasts will generally warm by around 3°C (5.4°F); temperate lands by more than 5°C (9°F); and the poles by as much as 8°C (14.4°F). As we have already seen, northern Canada, Alaska and Russia are already experiencing the effects of the warming of previously frozen soil, with thawing of the permafrost causing extensive damage to buildings and roads. The world will be not just warmer but wetter, with much of the increased precipitation falling as snow. On the other hand, subtropical areas will see a 30 per cent drop in rainfall, causing a vast swathe of land to dry out from Mediterranean Europe and northern Africa through the Middle East to central Asia. Another parched belt will cover southern Africa. These regions will increasingly see crippling droughts and famines as soil fertility deteriorates and crop yields decline. Although the rising carbon dioxide level will at first trigger more lush growth of crops such as rice and maize, once temperatures rise by more than 3–4°C (5.4–7.2°F), cereal production will start to fall off. In fact, if the warming exceeds 4°C (7.2°F), then entire regions, such as virtually all of Australia, may be too hot and dry to grow crops. Hike temperatures by more than 6°C (10.8°F) and farming is even threatened in temperate lands.

Meanwhile, higher latitudes are likely to get wetter as the air warms and storm tracks move. Hurricanes are likely to become more intense. Which all means that even in the relative comfort of temperate Europe, we will feel the bite of global warming. Here, winters are likely to get substantially rainier, with devastating river floods

predicted to become more regular across the continent. While cold-related deaths in northern Europe are expected to decline over the next century, severe droughts and heat in southern Europe might kill many tens of thousands of people every year. The hot, dry conditions will mean that wildfires, which have been terrible in countries such as Portugal in recent years, will get even worse. A temperature rise of only 3°C (5.4°F) could give Mediterranean Europe catastrophic droughts once every decade, causing water shortages for billions. And those in the north won't escape the hot weather either. The summer of 2003 was the hottest northern Europe had experienced in 500 years, with temperatures just 2–3°C (3.6–5.4°F) warmer than usual triggering a heatwave that killed 25,000 people and cost $15 billion in losses to forestry and farming. By the middle of the century, the 2003 scorcher could be a typical Parisian or London summer.

All in all, Europe's long-term forecast is for more extreme weather all round. And this is likely to have a significant impact on our modern lifestyle. Every year, northern Europeans flock to the Mediterranean in what is the single largest flow of tourists across the globe, accounting for one-sixth of all tourist trips in 2000. It is an annual migration of 100 million people who spend an estimated total of £67 billion per year, but it won't last. In the future, as southern Europe and northern Africa become parched and scorched, the package tourists will desert the Mediterranean resorts. They won't switch to cooler skiing holidays among Alpine peaks because the snow and ice cover there is disappearing, threatening many mountain resorts too. Unlikely as it might seem, future generations of holidaymakers may instead switch to the increasingly balmy Baltic beaches of Tallin, Riga or Gdansk.

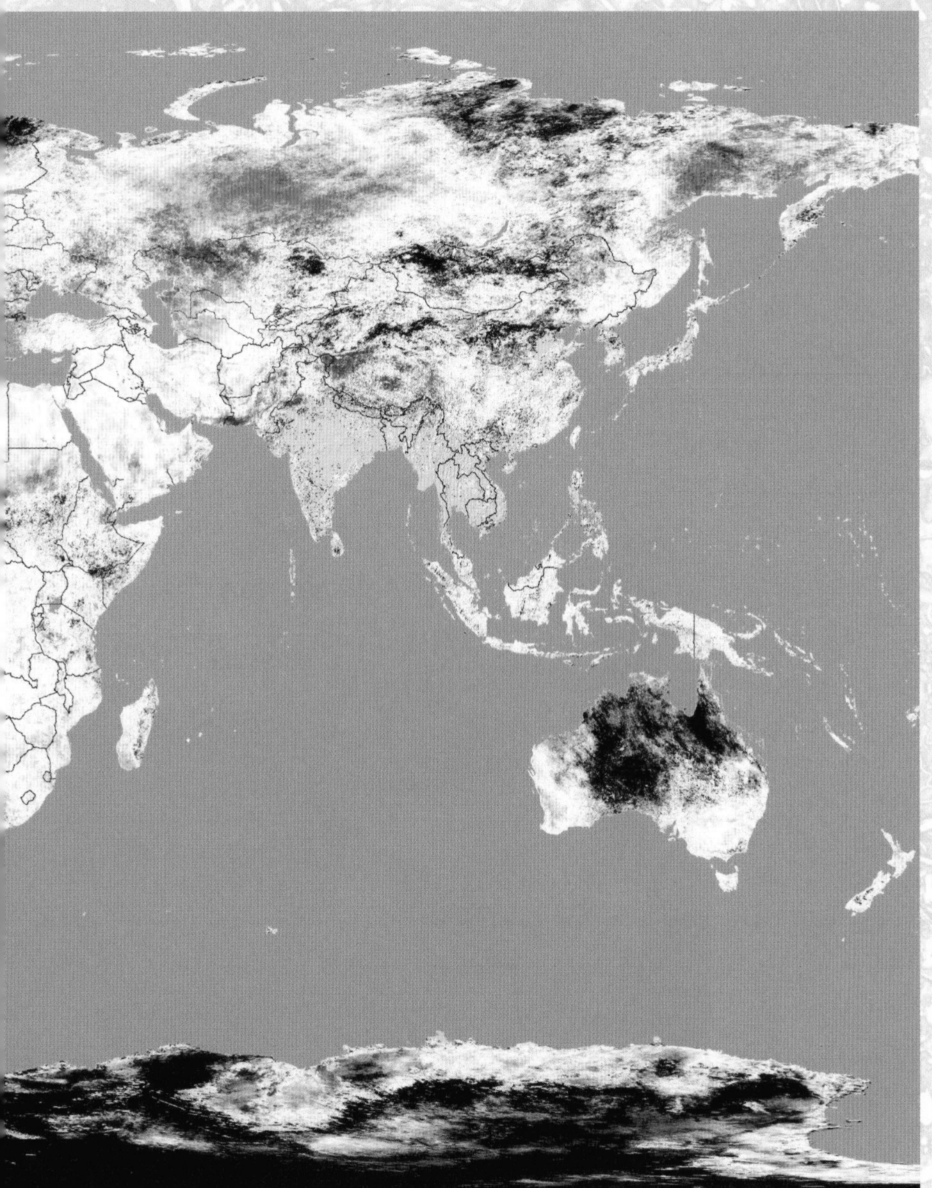

LEFT **THE HEATWAVE OF JULY 2006 WAS NOT CONFINED TO EUROPE – MANY OTHER PARTS OF THE WORLD EXPERIENCED THEIR HOTTEST JULY ON RECORD. THE MAP REVEALS WHERE TEMPERATURES WERE HIGHER (RED) OR LOWER (BLUE) THAN THE JULY AVERAGE OVER THE PREVIOUS FIVE YEARS, WITH DARK RED DENOTING TEMPERATURES 10°C (18°F) ABOVE THE NORM.**

ecological change are also apparent. For example, many species have been moving poleward by an average 6 km (4 miles) per decade for the last 30–40 years, and seasonal events such as flowering and egg laying appear to be occurring 2–3 days earlier each decade in the temperate lands of the northern hemisphere. Nothing but global warming seems capable of explaining all these effects.

It looks pretty bleak for our future (*see* 'Prognosis for a Planet', page 18), but some words of caution are needed regarding these gloomy scenarios. Most of the forecasts are based on sophisticated computer simulations of Earth's climate, but this is a science that is still finding its feet. It is only since 1985 that global climate models have been at all credible, and only since 1989 have there been supercomputers powerful enough to start simulating the cascade of knock-on effects that bounce back and forth between land, ocean and atmosphere. Around 3 billion dollars is spent on research into climate change every year, much of it trying to simulate how various cogs and wheels will operate under countless possible future scenarios. These models do not tell us the future – they tell us a range of possible futures. And they give us an extraordinary insight into the wondrous complexity of our planet's workings, revealing how it might react to change as a single, self-regulating system – almost as a living creature in its own right. More importantly, the models reveal the weak links in Earth machinery, the danger zones where climate change could cause vital systems to start to break down. These are, if you like, the Achilles heels in Earth's defences. We've already met one of these Achilles heels – the peatlands of Siberia – and in the next few pages we'll introduce some others, and show how the intricate way in which our planet works is both reassuringly robust and alarmingly unpredictable.

RIGHT **THE LEGENDARY DUST STORMS OF THE SAHARA ARE A VITAL LINK IN OUR INTERCONNECTED WORLD – THE WIND-BORNE DUST BLOWS FAR FROM THE CONTINENT, FERTILIZING FORESTS AS FAR AWAY AS THE AMAZON BASIN.**

Poisoned Lungs

The Amazon rainforest is Earth's greatest celebration of life, home to a third of the planet's land species, from insects to jaguars and everything in between. But for all the splendour and diversity of the Amazon's animals, it is the plants that constitute the true natural wonder. The world's rainforests are often described as our planet's lungs, but they are not: they consume as much oxygen as they release. They are, if you like, oxygen-neutral. Nevertheless, forests worldwide absorb some 25–30 per cent of the carbon dioxide in the atmosphere, and the Amazon is one of the main consumers. As human emissions push carbon dioxide levels ever higher, tropical rainforests will initially flourish, much like plants thriving in a greenhouse. But the Amazon's love affair with global warming is expected to be short-lived – beyond a certain tipping point, a fraction of a degree more warming could kill it.

That's because the Amazon basin may be on the brink of catastrophe. It currently covers a region nearly as big as the continental United States, but every year an estimated 15,000 square km (5800 square miles) of Amazonian forest – an area roughly the size of Wales – is burned or felled to make way for cattle ranching, farming and other development. What's more, selective logging – the supposedly environmentally benign practice of removing one or two trees and leaving the rest intact – might conceivably double that figure. And as humans hack down or burn the rainforest, the decomposing woody debris, roots and leaf litter left behind is releasing this store of carbon dioxide back into the air. Today, deforestation in the Amazon adds as much as 500 million tonnes of carbon to the atmosphere every year. As the planet's temperature rises, so the micro-organisms that consume forests will become ever more insatiable. Many scientists believe that, over the next 50 years or so, the Amazon rainforest will switch from being a major consumer of carbon dioxide to a significant producer of the stuff. And when that happens, the Amazon's copious emissions will make the world greenhouse even hotter. A change in just one part of the global system will have as yet unforeseen consequences elsewhere.

For a long while, this worry has been tempered by the belief that the Amazon rainforest is robust enough to cope with such changes. After all, it has been valiantly recycling our atmosphere ever since the excesses of the methane outburst 55 million years ago, and during the drier climate of recent ice ages the forest was thought to have retreated into isolated patches, or 'refugia', where new species evolved in isolation before bursting out again in warmer periods to restock the natural ark. But new research shows that the Amazon rainforest endured the arid cold spells of ice-age Earth pretty much intact. Rather than being an old hand at bouncing back from natural deforestation, the Amazon now appears to be completely unprepared for the unprecedented changes currently being inflicted on it. The most extreme scenarios suggest that half this pristine wilderness will be lost in the next 50 years. A forest that has survived for tens of millions of years could be irreparably damaged within a century.

Shifting Sands

The sandy winds of the Sahara desert are spectacular and legendary: winds like the haboob, a Sudanian dust storm that creates a bright yellow wall 3000 metres (10,000 feet) high and is followed by torrential downpours. Or the simoon, which the ancient Greek writer Herodotus records as engulfing whole armies; one nation, he wrote, was 'so enraged by this evil wind that they declared war on it and marched out in full battle array, only to be rapidly and completely interred'. Or the harmattan, a 'sea of darkness' that blasts west across the Sahara as a red fog to deluge lands as far away as England in showers of mud so crimson that they have been mistaken for blood. The harmattan blows west, its gusts dissipating over the Atlantic Ocean, but its enormous cargo of red dust – millions of tonnes of earth – sweeps on, carried high in the atmosphere, to eventually

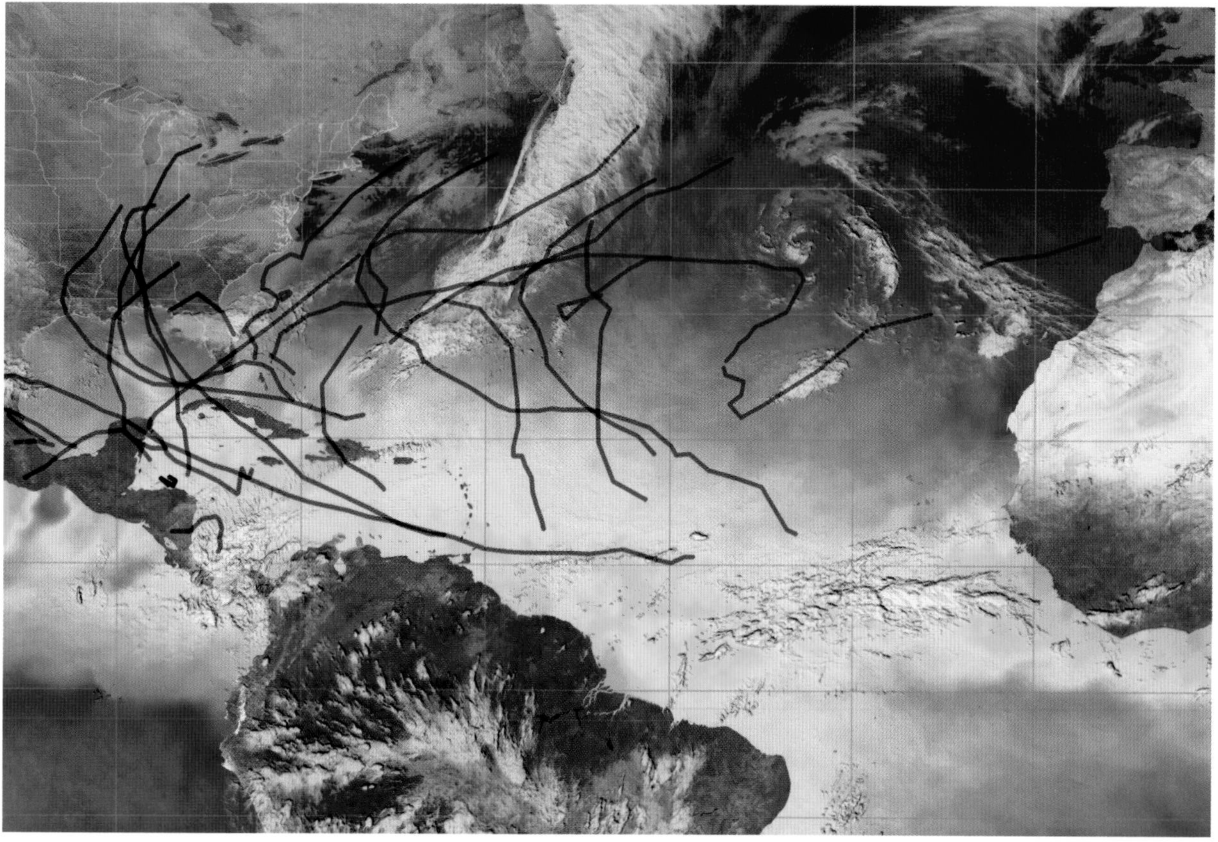

ABOVE **HURRICANE TRACKS FOR THE 2005 SEASON SHOWING THEIR SPAWNING GROUNDS IN THE WARM TROPICAL WATERS OF THE ATLANTIC. THE 2005 HURRICANE SEASON BROKE RECORD AFTER RECORD; MOST NAMED STORMS, LOWEST PRESSURE MEASURED IN THE ATLANTIC, LONGEST-LIVED DECEMBER HURRICANE... THE LIST GOES ON.**

rain down on the Amazon basin. It is this continual delivery of powdered minerals that fertilizes the Amazon basin. For thousands of years this transoceanic nutrient supply has enriched the rainforest, increasing its abundance of life. But, with global warming, that may be about to change, for the great desert, it seems, is more fickle than we might imagine.

The Sahara desert seems such a vast centrepiece of the world map that it comes as a surprise to discover that it was born only 6000 years ago. A small change in the distribution of incoming solar radiation, due to a subtle change in Earth's orbit, had weakened the equatorial storms that fed the African monsoon. Within a few decades, the tropical summer rains that once watered much of northern Africa had retreated south, and vast areas of woodland and marsh had become parched wasteland. Over the following centuries, the drifting sands of the desert spread north as well, and the ancient peoples who had farmed the once fertile Saharan heartland were pushed out. Part of the exodus moved east to settle a river valley that had previously been too marshy, and so began the Nile civilization and the age of the pharoahs. Others remained in isolated havens where water was still available, but by 2000 years ago only one group of hardy people was left holding back the desert: the Garmanthians, skilled charioteers who held in check the southward advances of imperial Rome. But on their other flank, the advance

ABOVE **VENICE IS ONE OF NUMEROUS LOW-LYING COASTAL CITIES FACING PERMANENT INUNDATION FROM RISING SEAS OVER THE NEXT CENTURY.**

of the desert was unstoppable. By AD 500, the Garamanthian culture was gone, its people scattered to a nomadic existence and its ruins buried beneath the sand. A tiny change for the planet had played out with dramatic effects at the scale of human history.

Today, with its mountainous dunes and vast sand seas, the Sahara is perhaps the archetypal desert. But climate models suggest that it may be on the verge of a sudden flowering. Although its northern portions look likely to become steadily drier, a warming of the tropics ought to strengthen the African summer monsoon, so in decades to come the Sahel and Saharan interior may once again receive regular rain. Rain would allow the growth of vegetation, which, in turn, would absorb moisture and humidify the air, stimulating more rain – a self-sustaining cycle of renewal. But every upside has a downside on the interconnected globe, and as plants consolidate their grip on the loose ground, their roots gradually stabilizing the once friable earth, the Saharan winds will be robbed of their dust, and across the ocean the Amazon rainforest will pay the price.

The winds of the Sahara might have another sting

in their tail. Years in which powerful dust storms have swept across the ocean have seen fewer hurricanes batter the Caribbean and the Gulf of Mexico. There is reason to believe that African dust clouds might have a dampening effect on these terrifying storm systems. Some of the worst hurricanes have happened when the air was relatively dust-free, so the dust appears to be a factor, although the ferocious energy of hurricanes is fed not by the wind but by the sea, which plays a leading role in the distribution of heat energy around our planet.

Meltdown

The warm waters that swirl through the Gulf of Mexico do not just feed hurricanes with energy. They sweep north from the equator and flow across the Atlantic towards western Europe and the Arctic, tempering the local climate in the process. This so-called Gulf Stream is just one stretch of a global ocean current that flows around the planet like a conveyor belt, exporting heat from the tropics and redistributing it. All the climate models indicate that when the ocean conveyor slows, the world cools. And if it were to stop, then the planet would catch a nasty cold. So reports that waters of the North Atlantic are getting less salty are causing some ocean scientists to worry that melting Greenland and Arctic ice is already making the ocean conveyor sluggish. There are also signs that the northward transport of heat in the North Atlantic has decreased by as much as 30 per cent in the last 50 years. For those of us in Europe, such a go-slow might initially cool the balmy temperatures expected from future global warming, but it could leave the tropics and the southern hemisphere sweltering. And, since vigorous ocean currents are needed to remove large amounts of carbon dioxide and heat from the air, in the long run the short-lived cold snap would probably accelerate global warming.

Studies of how the last ice age stuttered to a close suggest that what could slow the ocean conveyor belt would be an influx of fresh water into the Atlantic Ocean. Some 12,000 years ago, a sudden and massive surge of meltwater from a decaying ice sheet, which had once kept eastern Canada, New England and much of the Midwest under several kilometres of ice, swept into the North Atlantic and in doing so kick-started a dramatic fall in global temperatures. A world slowly creeping out of the ice age was abruptly sent back into the freezer for a thousand years – a last hurrah for ice-world Earth. Today, the vast meltwater lakes are all gone from North America, but the Arctic still has one gigantic reservoir of fresh water left: the Greenland ice sheet.

During the 10,000 years that have passed since Earth was last held in thrall by ice, this humungous mass of recrystallized snow has remained virtually unaltered, and geologists have long assumed that it would take centuries or even millennia for it to melt. But now some scientists believe the gargantuan glacier is close to a tipping point that might induce its total meltdown. The devil, however, is in the detail, since each season's measurements reveal strong variations in the temperature of the air and the ice, the amount of meltwater generated, and the rate at which the great ice mass itself is moving. Such variability is typical of complex natural systems and can make it difficult (and at times foolhardy) to try to identify meaningful trends and tendencies. Nevertheless, a clear and consistent signal is emerging. As we'll discover in a later chapter, there are signs that Greenland's ice is changing fast.

Some climate models predict that a warming of less than 3°C (5.4°F) – likely in this part of the Arctic within a couple of decades – could start a runaway melting that will eventually raise sea levels worldwide by enough to inundate the land on which more than 430 million people depend. Half of Florida would be drowned, small Pacific and Caribbean islands would disappear, and low-lying coastal cities like Tokyo, Shanghai, Hong Kong, Mumbai, Kolkata, Karachi, Buenos Aires, St Petersburg, New York and London would be waist-deep or knee-deep in water.

An even larger reservoir of water lies in Antarctica, where more than 60 per cent of the world's fresh water is trapped as ice. If it were to melt, sea levels could rise by 60 metres (200 feet). The vast East Antarctic ice sheet is unlikely to break up soon, but the smaller West Antarctic ice sheet is already causing alarm. In 2002, a shelf of sea ice the size of Rhode Island disintegrated on the West Antarctic coast, opening up a bay that had probably been ice-bound for 12,000 years. And there are worries that even parts of the ice sheet anchored to solid rock could swiftly become unstable, since much of the West Antarctic bedrock is below sea level. If this slumbering giant really has woken up, it could push the planet's climate over a precarious tipping point.

Global Village

In 2007, for the first time in history, the world's growing urban population outnumbered the rural population. For thousands of years, the size of cities was limited by disease, but with the advent of improved medicine in the eighteenth century, cities expanded rapidly, their populations swelled by an influx of rural migrants and a new imbalance between birth and death rates. 'Supercities' – those with a population exceeding 2 million – first developed in the late nineteenth century, and by 1950 the planet boasted its first two 'megacities', London and New York, with populations exceeding 8 million. Just over half a century on, we now have more than 25 megacities and several hundred supercities. And these vast urban sprawls are growing larger by the day. Their inexorable expansion has made it necessary to invent a new term, 'metacity' (though some prefer 'hypercity'), for conurbations with more than 20 million people – the entire population of the world at the time of the French Revolution. Today only metropolitan Tokyo has incontestably passed this threshold (although Mexico City, Seoul and New York make some lists), but by 2025 Asia alone might have ten or

RIGHT **THE METROPOLITAN DISTRICT OF GREATER TOKYO REQUIRES FOR ITS SUSTENANCE A BIOLOGICALLY PRODUCTIVE LAND AREA MORE THAN THREE TIMES THE SIZE OF JAPAN.**

eleven such megalopolises, and similarly stellar growth is predicted for West Africa. In two decades time, more than 5500 million people will live in cities – more than our entire 1990 rural and urban population combined.

The growth of these urban leviathans is a new experiment for life on Earth, for our swelling 'global village' has a ravenous appetite for land and resources. Greater Tokyo, for example, requires for its sustenance a biologically productive land area more than three times the size of Japan. The West is no less demanding. It is said that if everyone on the planet were to consume natural resources and generate carbon dioxide at the rate we do in the UK or USA, then we'd need two more planets to support us. To satisfy these cravings, the human race, for the first time in its history, now moves more rock and soil than all of nature itself. Through the bulldozer and chainsaw, half of Earth's land surface has been domesticated for direct human use since the beginning of the Industrial Revolution. In the same three centuries or so, the world's population has increased tenfold to 6000 million – a population explosion that is straining our planet's capacity to cope. And as our cities expand to vast proportions, so does the potential scale of the natural disasters that our unpredictable planet will continue to fling at us. Direct urban hits from future seismic and volcanic spasms of the not-so-solid Earth (or modest strikes from space debris) are likely to produce death tolls of hundreds of thousands or even millions, and storms, floods and chronic droughts will have the capacity to cripple tens of millions. And, of course, against this backdrop of Earth's natural capriciousness is the spectre of society's wilful chemistry experiment with our planet's defences.

OPPOSITE **THE SHEER ENORMITY OF HUMAN MODIFICATION OF THE EARTH, SUCH AS HERE AT THE SUPERPIT GOLDMINE IN KALGOORLIE, WESTERN AUSTRALIA, HAS LED MANY TO ARGUE THAT WE HAVE ENTERED A NEW GEOLOGICAL AGE: THE AGE OF HUMANKIND (THE 'ANTHROPOCENE').**

The Road Ahead

It seems clear that our planet is undergoing extraordinary changes, but what exactly do these changes mean for our future world? Are they excessive when compared with Earth's erratic behaviour over the last half a million years, when ice sheets repeatedly advanced and retreated across much of the planet? Or when viewed in the light of hundreds of millions of years of continual geological change?

In this book, we will show how our planet acts like a living, breathing organism – one that regulates its temperature, burns energy, continually renews its skin, and has a face that changes with age. We will tell the story of its life: how it was born and will one day die, and how it has been maintained by a remarkable circulatory system. Each chapter will look at one aspect of Earth's 'metabolism' – impacts from space, the heat engine within, the atmosphere, the oceans, and ice – and explore the critical role it plays in keeping the planet alive.

The result is a remarkable tale of survival, revealing a planet that is phenomenally tough and has the power to heal itself. Even when cataclysmic events caused Earth's life-support mechanisms to falter, plunging the world into the deep freeze or turning it into a hothouse, Earth survived. In fact, these near-death experiences seem to have been great turning points. In spite of – or perhaps because of – such traumatic brushes with catastrophe, the planet became a home for life – our home.

But that home would now seem to be at risk once again. The human race appears to be forcing changes that are straining our planet's defences. This fragility seems curiously at odds with Earth's long-standing resilience. Daily we are bombarded from all quarters about the need to 'save the planet', but in this book we pose an even more fundamental question: is it the planet that needs saving from us or is it the other way round?

CHAPTER ONE

Impact

We live on a rock-coated metal ball hurtling at 107,000 km/h (66,000 mph) through space. It's an unsettling thought not made any less so by the knowledge that the space we're careering through is far from empty. Like a car speeding through congested city streets, our breakneck orbit around the Sun takes us across the path of millions of projectiles – everything from tiny pebbles to rocks the size of the Ukraine. Earth has been cruising through this galactic gauntlet for the last 4.5 billion years (an unfathomable timespan to which we'll return later), so it is not surprising that along the way it has taken a few bumps, and even suffered the odd head-on collision. It turns out that such cosmic pile-ups lie at the very heart of our planet's story. Encounters of the extraterrestrial kind, we will see, have been a driving force for planetary change since Earth's very birth, instrumental in making Earth habitable and giving us the ingredients for life – perhaps even life itself. Yet few of us looking to the heavens sense a planet whirling through celestial flak, with us anchored to it by the invisible tug of gravity. Under a starry sky with its familiar constellations, our little blue dot in space seems serenely isolated from the whizzes and bangs of a violent and capricious universe. But every so often, planetary violence comes to town.

OPPOSITE **COMETS, SUCH AS THIS ONE SEARING THE NIGHT SKY IN MARCH 1996, ARE GIANT ICEBALLS FROM THE FAR-FLUNG RECESSES OF OUR SOLAR SYSTEM WHICH HAVE DELIVERED WATER, HEAT AND THE ODD MOMENT OF DEVASTATION TO OUR ROCKY HOME.**

On 16 July 1994 – the 25th anniversary of the launch of *Apollo 11*, the first manned lunar landing mission – comet Shoemaker-Levy 9 smashed into the planet Jupiter. The collision was entirely foreseen, the comet having been spotted the previous year by the comet-hunting team of Gene Shoemaker, his wife Carolyn and colleague David Levy. Some 60 years earlier, this icy body from the outer reaches of space had been captured into orbit around Jupiter and began slowly spiralling in to its imminent destruction. But the nature of the collision was a surprise. For a start, it didn't hit as one big lump. As Shoemaker-Levy 9 drew close to Jupiter, the giant planet's phenomenal gravitational pull ripped the comet into a string of about 25 enormous chunks from 300 metres (1000 ft) to 2 km (1.2 miles) wide. Over the course of the next few days, astronomers at hundreds of observatories around the world watched as the fragments ploughed, one by one, into Jupiter's southern hemisphere. Orbiting Earth, the Hubble Space Telescope had a grandstand view of the carnage as fragment after fragment pounded into the thick Jovian atmosphere, each creating giant fireballs that sent plumes of debris rising thousands of kilometres above the planet's cloud tops and left immense, dark bruises in Jupiter's stratosphere. The colossal scale of the impacts astounded scientists. As one astronomer put it later, 'We were thinking that we were going to have to go in with a microscope ... but it's just blasting away at us ... unbelievable.' The largest strike was from a fragment known as nucleus G, which hit with the force of 6 million megatonnes of TNT – 75 times all the nuclear weaponry in existence. Nucleus G was only the size of a small mountain but it threw up fireballs as big as Earth.

For those watching, the spectacle of Shoemaker-Levy 9's shock-and-awe assault on Jupiter was a sharp reality check. Until that moment, our corner of space had seemed a safe and tranquil haven, but now Earth looked exposed and vulnerable. If the comet had struck us rather than Jupiter, it would have been a civilization-ending event. Instead, from our safe vantage point 42 million kilometres (26 million miles) away, we enjoyed a live demonstration that the universe is indeed a violent place.

A Violent Birth

It was long thought that the birth of Earth and its terrestrial neighbours – Mercury, Venus and Mars – was a regular and tidy affair, the ordered emergence of solid bodies condensing gradually out of a hot, swirling disc of gas and dust that encircled a young Sun. Within this 'solar nebula' there were occasional bruising encounters between emerging planets and wandering debris, but in general it was a relatively painless birthing process. However, it now looks as though chaos reigned in those early years, with random collisions being the dominant process by which the inner planets (and perhaps even the outer planets) came to be. According to this 'planetesimal theory', grains of dust within the turbulent veil of the solar nebula stuck together to form clumps; the clumps clustered into metre-sized rocky lumps; and these in turn coagulated into kilometre-sized bodies – planetesimals. In time, a few of the planetesimals grew to be hundreds or thousands of kilometres wide and so became planets. This was planetary cannibalism, as the growing bodies crashed into and consumed the flotsam and jetsam around them.

Computer simulations suggest that the rocky planets of the inner solar system (Mercury, Venus, Earth and Mars) formed in a feeding frenzy in which around a hundred Moon-sized bodies, ten Mercury-sized bodies and several Mars-sized bodies were devoured. Around half to three-quarters of Earth's present mass was accreted from massive bodies that would have made respectable planets in their own right if they hadn't succumbed to Earth's growing appetite. Indeed,

OPPOSITE **THE DARK SMUDGE ON JUPITER'S SWIRLING ATMOSPHERE IS ONE OF A STRING OF GIGANTIC DEBRIS PLUMES THROWN UP BY THE IMPACT TRAIN OF COMET SHOEMAKER-LEVY 9 IN 1994.**

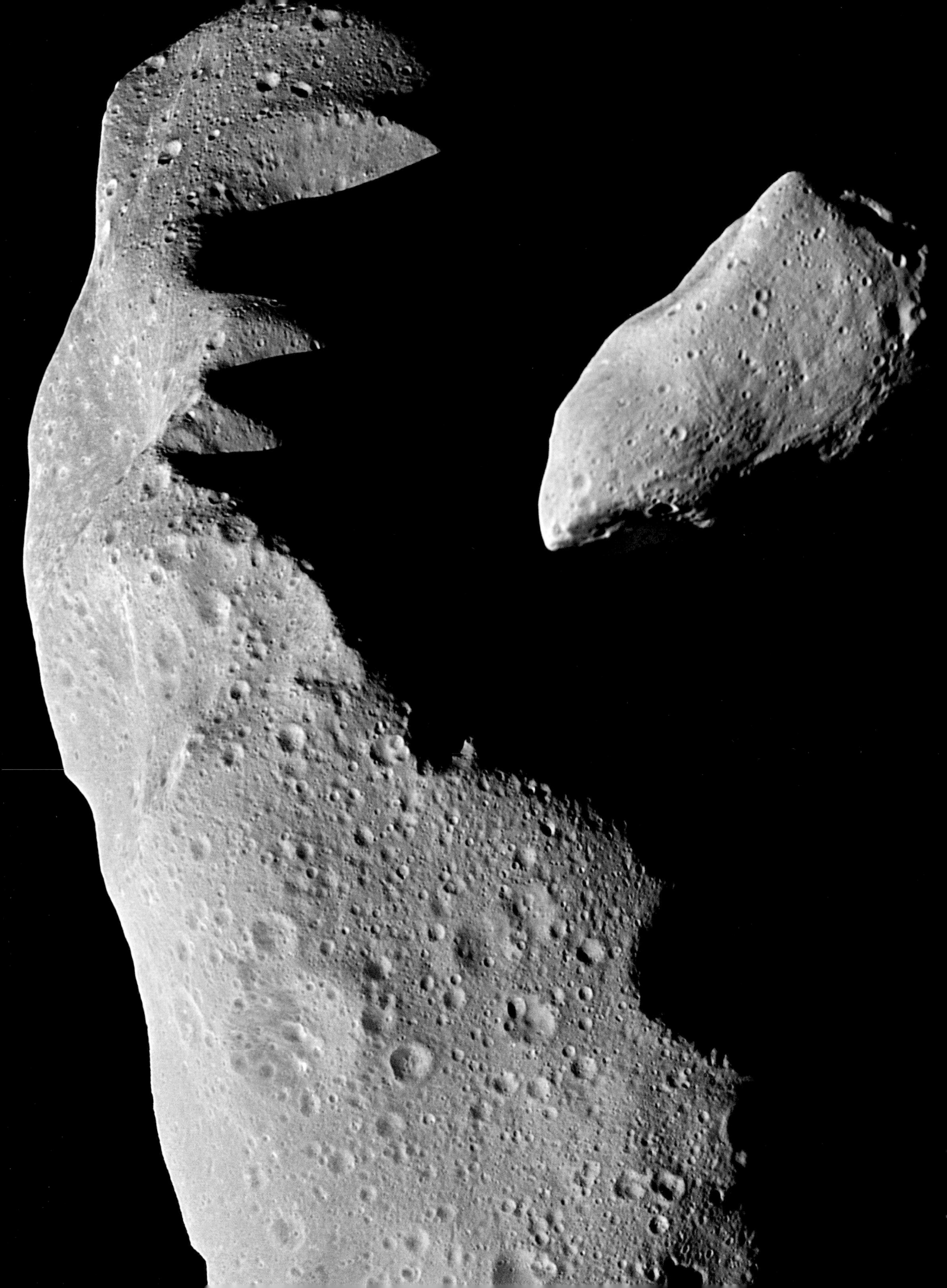

many of them already had a fully formed metal core and a rocky outer carapace. When they struck Earth, so much heat was released that widespread melting occurred as they broke up, allowing fragments of opposing metal cores to fuse and readily marrying the colliding rocky mantles. Earth may therefore have grown quickly by piecemeal assembly of large, prepackaged planetary morsels. The appetite of the expanding Earth and its three similarly swelling neighbours was quelled only when their feeding zone had been sucked clean of debris, leaving the inner solar system virtually clear of planetesimals.

OPPOSITE **ASTEROIDS IDA (LARGE) AND GASPRA (SMALL), PHOTOGRAPHED TWICE BY THE *GALILEO* SPACECRAFT IN AUGUST 1993.**
ABOVE **THE ROBOT SPACECRAFT *NEAR-SHOEMAKER* PHOTOGRAPHED ASTEROID EROS BEFORE CRASHING INTO IT IN FEBRUARY 2001.**

But there was plenty of planetary detritus left elsewhere. By convention, the large chunks of rocky debris that orbit the Sun mostly outside the inner solar system are known as asteroids. These are pebbles and boulders that never quite made it into planets, or, if they had done, were smashed off again to return to their itinerant ways. The gigantic planet Jupiter, which has more than twice the mass of all the other planets combined, stands guard over these rogue boulders. The Jovian giant's bulk – 318 times that of Earth – acts as a massive gravitational magnet, luring peripatetic chunks of rock and metal away from a collision course with the inner solar system. Much of the debris gets trapped in what is known as the Main Asteroid Belt, a wilderness area where stray asteroids roam in their native habitat. Something like a million rocky objects more than 1 km (0.6 miles) wide are pooled here. Jupiter's protective influence ensures that Earth is shielded from most of this freewheeling debris, but despite the best efforts of our planet's minder some slip through the net. Every few thousand years, a gravitational tug-of-war between Jupiter and Saturn gives a few asteroids enough momentum to break the Jovian grasp, and they set off on new orbits, some destined to cross our path. Most plunge harmlessly into the Sun, but roughly a third of these Earth-crossers will crash into our planet.

Jupiter itself lies close to the 'snow line', an invisible divide in the solar system that separates the dust, rock and metal bodies of the inner solar system from the gas- and ice-rich material of the outer solar system. Beyond the snow line, water (the combination of hydrogen and oxygen that makes the most common molecule in the universe) exists mainly in the form of snow. It was by gobbling up gargantuan amounts of water that Jupiter and its portly neighbour Saturn swelled into huge, gaseous, stormy worlds. Beyond these so-called gas giants, in the even colder reaches of space, lie the icy worlds of Uranus and Neptune, and the recently downgraded mini-planet Pluto. Out in these frigid recesses of our solar system, the builders' rubble left over from the birth of the planets occurs as enormous lumps of snow, frozen gas and dust. These icy wanderers present the other great threat to planet Earth: comets.

Comets

Millions of comets linger in the 'Kuiper belt', a dark and mysterious zone beyond Neptune, but billions inhabit the even more far-flung outpost of the 'Oort cloud', a spherical halo of icy debris beyond the limits of our solar system. They are often immense – many hundreds of kilometres in diameter – and from their distant spawning grounds, long, looping trajectories freewheel these giant snowballs into the heart of our solar system. Each

time they skirt the Sun, a hurricane of solar particles – the solar wind – evaporates some of their gas-rich outer layers. The resulting vapour trails give comets their distinctive tail of light. After repeated solar fly-bys, a comet's snowy armour may vaporize entirely, leaving a nucleus of rubble that looks much like an asteroid. But comets travel at up to 70 km per second (43 miles per second) – three times faster than a typical asteroid – so they have the potential to strike a planet with enormous energy. Which is why Shoemaker-Levy 9 made such a spectacular mess of Jupiter.

Seen as harbingers of doom since ancient times, comets have long been a source of fear and superstition, held responsible for everything from the paranoia of the Roman emperor Nero to the collapse of the Aztec empire. Even today they are the focus of religious revivals and mass suicides. Many comets, especially those spawned 'nearby' in the Kuiper belt, take anything from a few decades to two centuries to circle the Sun and so are regular visitors to our neighbourhood. In fact, their paths have been so well studied that we know these frequent fliers pose little foreseeable danger to Earth. Harder to gauge are the habits of those Oort comets that come from far beyond the solar system. Knocked out of their remote orbits by the gravitational tug of gas clouds in deep space, they career towards us on wildly elliptical orbits that sweep close to the Sun. Some take an incredibly long time to complete their orbits, which means that, in terms of human history, they are single-pass events. These one-hit wonders emerge from the blind side of space unannounced and travel so fast that they give a forewarning of only months. What's more, they can arrive on our doorstep immense because they haven't been winnowed by repeated solar fly-bys. Comet Hale-Bopp, prominent in the sky in 1997, had a girth of at least 25 km (15 miles)

LEFT **COMET HALE-BOPP'S BLAZING VAPOUR TRAIL LIGHTS UP THE JAGGED PEAKS OF MOUNT WHITNEY IN CALIFORNIA'S SIERRA NEVADA MOUNTAINS IN 1997.**

THE HABITABLE ZONE

In the recent Hollywood blockbuster *Sunshine*, the dwindling power of the Sun's nuclear fires threatens to plunge Earth into a deadly new ice age. To save the human race, a spaceship is dispatched sunward to rekindle our star with a nuclear payload the size of Manhattan. It might seem churlish to point out that as the Sun burns out, it will actually get hotter, or to note that since the Sun is powered by the equivalent of 4 billion hydrogen bombs exploding every second, our neighbourly delivery might not quite do the job. But what the movie does highlight is how absolutely dependent Earth is on our star. Virtually all life on Earth owes its existence to the Sun's brilliance: its energy powers every plant and sustains nearly every animal. If we could harness the energy output of this solar powerhouse for a single second it would supply the world's energy requirements for a million years. That's because every second 5 million tonnes of the Sun is converted to pure energy. Thankfully for us, the Sun is only halfway through its fuel supplies, but later in the book we'll chart exactly what will happen when it finally runs out.

For now, just admire the beautiful sunrises and sunsets. And think: this is what the power of nuclear fusion looks like from 150 million km (93 million miles) away. Conveniently, we are just the right distance from this nuclear reactor for our world to sustain life. Any closer and we'd bake like Venus, whose surface is nearly hot enough to glow; any further and we'd freeze like Mars, which is frozen to a depth of many kilometres. Astronomers call our narrow band of space the habitable zone, and if it weren't for the remarkable fact that Earth has remained within the zone's limits for the last 4.5 billion years, we wouldn't exist. That and the fact that our Sun is also just the right size. Suns can be all sorts of different sizes, but it turns out that ours is ideal. It's big – you could fit the Earth inside it a million times over – but if it were any bigger it would have raced through all its fuel supplies millions of years before complex life on Earth got going. If it were smaller, a wannabe habitable planet would have to be so close that it would risk getting stuck in a gravitational headlock, with only one side continually bathed in sunshine. Instead, we happen to have found ourselves at the ideal distance from an ideally sized star that has emitted an ideal amount of energy at a nearly constant rate for a very long time.

BELOW **JETS OF PLASMA (CHARGED GAS ATOMS) BURST FROM THE SUN'S SEARING SURFACE. SUCH OUTBURSTS CAN HEAD EARTHWARDS TO PLAY HAVOC WITH OUR COMMUNICATIONS. THIS IMAGE WAS CREATED BY CAPTURING EXTREMELY HIGH-ENERGY ULTRAVIOLET LIGHT WAVES THAT ARE INVISIBLE TO THE HUMAN EYE AND DEPICTED HERE IN BLUE.**

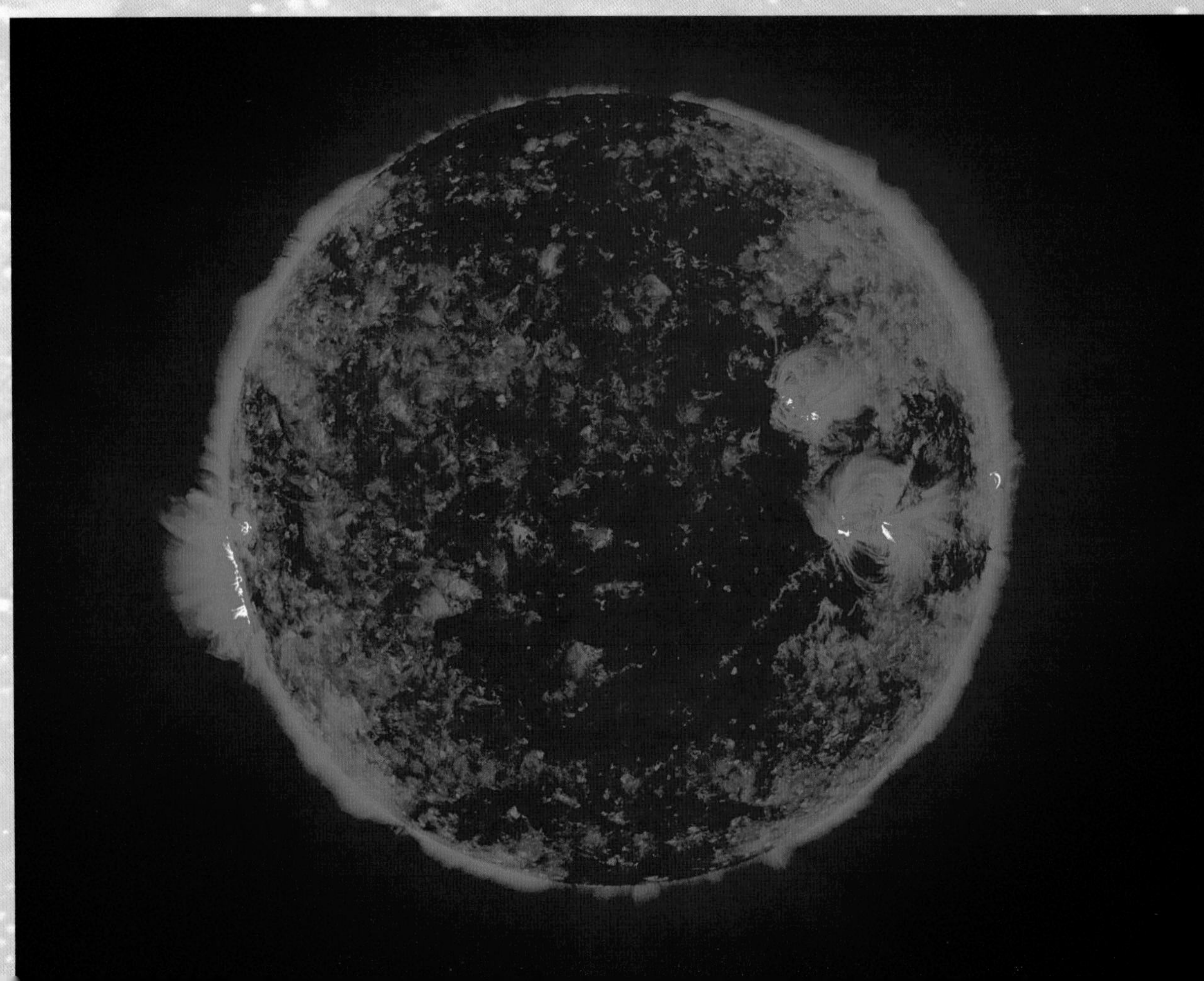

and possibly as much as 70 km (43 miles). It came within Earth's orbit, though thankfully on the other side of the solar system – which is just as well, because if this London-sized block of ice had struck, it could have sterilized our planet.

It's the stuff of Hollywood movies like *Deep Impact* or *Armageddon* – imminent and inescapable annihilation from space. So it is ironic that asteroids and comets have been a stabilizing, perhaps even life-giving, force during Earth's history. And in many respects, the story of Earth begins with its own big bang.

The Death of a Twin

According to our current view of the solar system, Earth and its neighbours are survivors of a random – and ultimately unique – barrage of collisions. The distinctive chemical make-up and orbital idiosyncrasies of each planet are legacies of its unique history of impacts. It was probably an impact with an object a fifth of its size that stripped Mercury of much of its rocky cover, leaving it with an apparently oversized core. Mars, it seems, owes the dramatic wobbling of its axis (the tilt of which varies from 0 to 60 degrees over a timescale of a few million years) to a collision, as does the weird, backwards rotation of Venus. Even the outer planets – the gas giants Saturn and Jupiter and the icy worlds of Uranus and Neptune – show signs of a violent history, with huge collisions probably responsible for knocking Uranus on its side and tilting Neptune. Among the wild wobbles and spins that arose from this planetary pinball, Earth's modest but stable tilt stands out. But, remarkably, that too is the consequence of a collision.

Once upon a time, our planet had a twin. 'Theia' is believed to have been about half as wide as Earth (making it roughly the size of Mars), and for several million years both planets appear to have shared a similar orbit around the Sun. But having two large planets so close meant that a clash was inevitable, and around 4.4 billion years ago their paths crossed. It was planetary fratricide: Theia was obliterated, most of its mass swallowed by Earth. The remaining pulverized debris spiralled into space, mixing with debris from Earth's shattered outer layer to form a cloud of rubble that eventually coagulated into our Moon.

The birth of our Moon may be the most singularly important event in the history of our planet. The Moon-forming impact was not the last major impact that Earth was to experience, but it seems certain to have been the last collision with a planet. Earth was now essentially fully formed.

The extra mass delivered by Theia gave Earth an instant bulking up, enlarging the core by some 20 per cent and giving the planet enough gravity overall to hold on to a substantial atmosphere. Most of the solar system's small rocky worlds haven't managed to hold on to an atmosphere, but Earth now had just the right mass and gravity to trap a blanket of water and gas around its surface. The collision would have stripped away our planet's primordial atmosphere, but a new one gradually emerged, this time held close by Earth's stronger gravitational pull.

And there were other dramatic consequences. When it first formed the Moon was much closer to Earth, appearing in the sky as a huge silver ball ten times bigger than today. Its gravitational tug made Earth spin much faster, resulting in a day being only five hours long. This, in turn, reduced the planet's previously erratic tilt, just as increasing the gyre of a spinning top will stop it tipping wildly. Were it not for the Moon, the tugs of the giant planets in our solar system would cause Earth's obliquity – the angle between the equator and the plane of its orbit – to vary by up to 80 degrees, periodically tipping the planet enough to make the poles sunnier than the tropics. Such wild climatic swings would have rendered the planet as uninhabitable as our madly tipping cousins Mars and Venus. Instead, the stabilizing tug of the Moon calmed the wayward excesses of the youthful Earth, leaving just a minor limp that we experience today as summer and

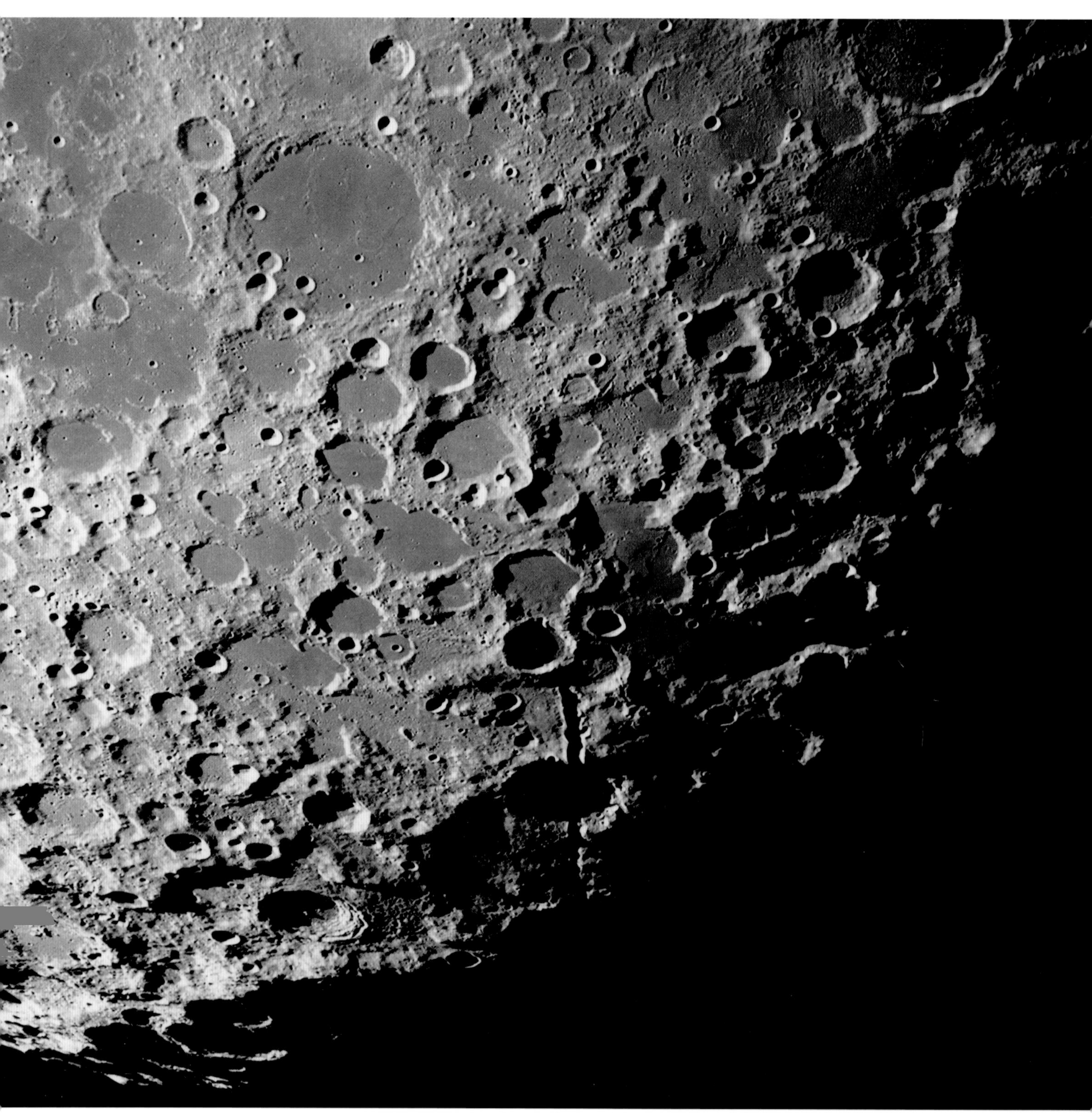

winter seasons. The Moon had become our planet's climate regulator.

The Moon's stabilizing influence gave Earth at least a chance of becoming habitable, but this was not the only benefit that our new satellite brought. With an Earth spinning much faster than today and a closer Moon in tow, you have the recipe for monumental tides. It is the gravitational pull of the Moon that pulls Earth's water out in a bulge and so gives us tides as the planet rotates. Four billion years ago, the close and rapidly orbiting Moon had a tidal pull that was truly immense. Four-hourly tides raised the sea level by 100 metres (330 ft) or so and exerted massive forces deep within our slowly congealing planet. These forces ripped open cracks in the newly formed ocean floor, creating new routes for heat to escape from below. The cracks became the site of hot springs, which, as we shall see in later chapters, may have been where life began.

Life may never have taken hold on Earth without the Moon's stabilizing influence and its tidal pull. Yet these life-giving effects are waning, because our next-door neighbour is slowly drifting away. The Moon has been getting further from us for billions of years and is currently receding by about 3.8 cm (1.5 inches) a year. This means that the erratic tilting that defined Earth's early days will gradually return, bringing catastrophic ice ages and ultimately ensuring a bleak future. For now, however, there is something more positive to reflect on. Due to an amazing stroke of luck, we happen to live at the precise point in Earth's history when the Moon's slow departure has made it appear exactly the same size as the Sun – and that's why total solar eclipses are so spectacular.

OPPOSITE **OUR MOON'S ANCIENT, HEAVILY CRATERED SURFACE PRESERVES A RECORD OF THE BOMBARDMENT THAT EARTH MAY HAVE ALSO ENDURED.**

Late Bombardment

As humans have peered into the dark recesses of space and probed and imaged their nearest planetary neighbours, a fantastic gamut of new exotic worlds has revealed itself. Most of these worlds are totally unlike the planet we call home, and in the next chapter we'll explore some of the features that make our world so distinctive. However, one thing common to all the planetary bodies thus far encountered is the sign of violent bombardment by space detritus. From Mercury, close to the Sun, out to the icy satellites of Uranus, impact scars abound. They range in size from tiny pits on lunar rock samples to giant, ringed basins more than 1000 km (620 miles) in diameter. All testify to skirmishes with now-vanished asteroids, comets and planetesimals. For the clearest evidence of this unceasing barrage we need only look to our next-door neighbour. With no air or water to erode them, the Moon's craters can remain pristine for billions of years – an enduring snapshot of a dangerous Cosmos.

After Galileo Galilei directed his telescope at the Moon in 1609 and recognized its circular features as depressions rather than mountains, debate raged as to whether lunar craters were formed by explosions below ground or impacts from above. The matter wasn't finally settled until the 1970s, when lunar rocks brought back to Earth by Apollo astronauts were all found to bear the hallmarks of collision. But those first Moon rocks had an even more intriguing tale to tell. They were all more or less the same age: somewhere between 4 billion and 3.8 billion years old. The Moon, it seemed, had suffered a short period of especially intense battering that astronomers dubbed the lunar cataclysm. The rocks had all come from the equator on the nearside of the Moon (where communication with the orbiting command module made it safest for the landing craft to be), so perhaps the samples just happened to be relics of the same impact. However, in 2000, scientists dated more than a dozen lunar meteorites that had been randomly blasted off the Moon before landing on Earth, and they found a similar cluster of ages. It looks as though the cataclysm, now renamed the late heavy bombardment, was a real and pan-lunar event.

And what happened to the Moon probably happened to Earth too. We are near enough to our lunar companion to have shared a similar history of assault. Except that, because our planet is ten times larger, we would have received roughly ten times more hits than our next-door neighbour. Even with the naked eye you can see evidence of the bombardment era on the Moon's face. The dark circles that we call lunar seas or maria are in fact gigantic impact basins – many the size of France – that filled with molten rock after being pounded 3.9 billion years ago in a frenzy of stupendous strikes. Their smooth surfaces bear surprisingly few craters compared to the paler lunar highlands, which are peppered with ancient impacts. And this tells us that after the burst of bombardment, it all went quiet.

Throughout our planetary neighbourhood from Mercury to Mars, there are signs of the renewed spate of violence between 4 billion and 3.8 billion years ago. Precisely what caused it isn't clear – perhaps the final repositioning of Uranus and Neptune in the outer solar system unleashed a storm of comets from the Kuiper belt. Whatever the cause, havoc reigned on the inner planets. If the Moon is anything to go by, then Earth would have had more than 20,000 craters the size of London or New York, about 40 basins the size of small countries, and several basins the size of entire continents. Major collisions probably slammed into us every century or so. As we'll discover later, there is no terrestrial sign of this fury – our planet has a unique ability to heal its scar tissue. But some scientists believe that the fact that the oldest known rocks on Earth date from around this extraordinary time implies that the

RIGHT **ALTHOUGH OUR MOON IS SLOWLY DRIFTING AWAY FROM US, ITS PRESENT DISTANCE AFFORDS US THE LUXURY OF ALMOST PERFECT SOLAR ECLIPSES LIKE THIS.**

Meteorites – Messengers from Outer Space

The Nullabor Plain of southern Australia is ideal meteorite-spotting country. Unobscured by scrub and free from human disturbance, the black, metallic blobs, burnished by their incandescent plummet through the atmosphere, are easy to spot against the arid red soil of the outback. Only the snow-white wastes of Antarctica make a better hunting ground. More than 1000 meteorites have been picked up in the Nullabor over the years, and the slow rate of weathering in the desert climate ensures they remain fresh-looking for thousands of years. Even lumps a couple of million years old still litter the ground in places. Spend long enough walking here and sooner or later you'll spot pieces of alien shrapnel – pebbles that hold ancient secrets about the universe.

Meteorites are the solar system's free samples, and they can tell us fascinating stories about our origins because they are the very stuff from which the planets were made. Some contain chemical elements in similar proportions to the Sun's atmosphere, and these 'chondritic' meteorites represent the most primitive material

ABOVE **IN METEORITES, THE SECRETS OF THE UNIVERSE ARE REVEALED.** BELOW **IAIN, METEORITE-HUNTING IN THE AUSTRALIAN OUTBACK.** OPPOSITE **THE IMILAC METEORITE – ALIEN SHRAPNEL FROM THE TIME OF OUR PLANET'S BIRTH 4.5 BILLION YEARS AGO.**

yet found in the solar system. By measuring the radioactive decay of their chemical elements, we can count back to when they formed. They turn out to be an almost inconceivable 4.6 billion years old (*see* 'How Old is Earth?', page 50). Others are equally ancient but have a different chemistry, courtesy of makeovers by heat and pressure during the violent era in which the planets were born. These 'differentiated' meteorites are priceless witnesses to the trauma in which our own planet came into being.

A few differentiated meteorites are much, much younger. These are chips of rock spalled off one of the planets and scattered across the solar system. Such a planetary diaspora cannot tell us much about the early days of the solar system, but it could hold clues about an even greater mystery: the origin of life.

On 28 June 1911, a perfectly innocent dog in Nakhla, Egypt, was minding its own business when it was flattened by one of 40 fragments of a large meteorite that had just exploded in the sky above it. What vaporized the world's unluckiest dog was a very rare type of rock indeed: one of only 13 Martian meteorites found on Earth. In 2006, NASA scientists broke open a piece of the precious Nakhla meteorite and discovered tiny veins riddled with carbon-rich debris. It looked like the work of bacteria. NASA had already reported controversial evidence of Martian microbes ten years earlier when scientists found strange, tube-like structures inside the endearingly named ALH84001, a meteorite found in Antarctica in 1984. The new discovery reopened the furious debate over whether extraterrestrial life had at last been found.

But what is certain is that even the most ancient chondritic meteorites can contain complex organic molecules, including more than 50 amino acids, of which eight are common protein-building types. Scientists now believe that interstellar space is replete with organic molecules, and that meteorites delivered these vital building blocks to the embryonic Earth. Whether or not life itself hitched a ride is still open to question, but if it did, there is every possibility that we are the offspring of Martians.

bombardment was sufficiently intense to melt and resurface our whole planet. Earth was essentially born anew.

The late heavy bombardment may have been the traumatic birthing process for life itself. For in ancient rocks that date from the aftermath of the bombardment, there are tantalizing hints of the first furtive biological activity. Traces of a particular type of carbon are thought by many scientists to be evidence of photosynthesis – the energy of life. It remains one of the most controversial areas of modern geology, but for some the first evidence of living things on planet Earth appears when the planetary blitzkrieg ends. Rich in the building blocks of life, the comets or asteroids that pummelled down are thought to have brought annual deliveries of 40 tonnes of amino acids and other organic molecules to the early Earth. What's more, as they smashed into Earth they released vast amounts of heat, perhaps kindling the first heat-loving organisms, thermophiles, which many biologists believe were the seedlings for the tree of life. We will take up this exhilarating story of the emergence of life in later chapters, but for now it is enough to note the irony that Earth's first simple inhabitants may have come out of the chaos of collision.

Shooting Stars

Today, 4 billion years later, it doesn't take much to appreciate that we live in a cosmic shooting gallery. Simply look at the heavens on a cloud-free, moonless night, preferably somewhere far from the glare of city lights. Watch for long enough, perhaps half an hour at most, and you'll see the exhilarating spectacle of a shooting star – a sudden but silent streak of light across the starry sky. Technically, though less evocatively, they are

LEFT FOR ALL THEIR SHEER SPECTACLE, METEOR STORMS, LIKE THESE CELESTIAL FIREWORKS PHOTOGRAPHED NORTH OF AYERS ROCK IN AUSTRALIA IN 2001, ARE GENERALLY HARMLESS, SHEDDING DEBRIS THAT MOSTLY REACHES THE GROUND ONLY AS SPACE DUST.

known as meteors, and they appear when fragments of shattered asteroids and/or dust particles shed by comets strike the outer edges of our atmosphere and burn up in sizzling fireballs. Barely 100 km (62 miles) of atmosphere is all that separates terra firma from outer space, but this thin air is enough to provide a pretty effective shield against incoming space debris. Travelling at thousands of kilometres an hour, meteors are rapidly incinerated by friction with air molecules and fleetingly glow with a brightness that can outshine the Sun. During these spectacularly incandescent falls to Earth, the braking effect of our atmosphere is so effective that the incoming rocky lumps rapidly disintegrate, spreading microscopic particles high in the stratosphere. Several tens of thousands of tonnes of this cosmic rain arrive this way each year, creating a dusty veil that swirls with Earth in its orbit and trails a wake. Over time, much of the meteor powder settles to Earth – so much, in fact, that every footstep a person takes contains a fragment of space dust.

Virtually all the fallout from meteor fireballs is too small to reach Earth's surface. So, for all their brilliance, these nightly firework displays have little real impact. When especially large or fast-moving meteors strike, a shower of rock and metal may reach the ground, but even when they crash-land as meteorites they are generally harmless. As yet, no-one is known to have been killed by a meteorite fall. Still, they arrive as a bit of a shock, as Arthur Pettifor found out when tending his onion patch in Glatton, England, on 5 May 1991: announced only by a loud whistling and whining, a dark stone the size of a grapefruit slammed into the garden just 20 metres (66 feet) from where he was standing. Or imagine the surprise of the residents of Barwell, England, when, on Christmas Eve 1965, an orange-red fireball screamed across the Leicestershire skies before breaking up above the small manufacturing town and spraying shrapnel into factories, houses, gardens, roads, and vehicles, amazingly causing no injuries. Around the globe, meteorites make a nuisance of themselves in this way, their destructive potential disarmed by the air above us. But every once in a while, something altogether deadlier manages to penetrate our planet's air defences.

Asteroid 'Air Bursts'

On 1 February 1994, as Shoemaker-Levy 9 was starting on its final descent towards Jupiter, a far more modest blast struck Earth's atmosphere. A military satellite had detected the brilliant flash of a nuclear explosion over Micronesia in the Pacific Ocean, and President Clinton, Vice President Gore and the US Joint Chiefs of Staff were woken from their sleep by Pentagon officials. Their fear was that a Chinese or Russian nuclear submarine might have accidentally detonated its nuclear bombs. Air force jets were dispatched to the area to examine the blast scene, but there was no sign of radiation. Scientific analysis revealed that the satellite had simply witnessed the explosion of an asteroid a few tens of metres wide. It had punched deep into the atmosphere, crossing the altitude at which airliners fly, before disintegrating midair in a violent blast – an 'air burst'. With the world not threatened with a nuclear incident, the president went back to bed.

Actually, space intruders like this are surprisingly frequent. Between 1975 and 1992, US Department of Defence surveillance satellites detected 136 large air bursts in the atmosphere. That's an average of about eight a year, but it's probably a gross underestimate because the spy satellites picked up the flashes purely by chance. Such midair meteor death throes radiate intense bursts of electromagnetic energy, as happened in January 2000 when the explosion of an object around 5 metres (16 feet) wide in the skies above Yukon in Canada triggered power cuts. If a similar-sized meteor exploded over London or New York, it could cause chaos. But what would happen if something bigger lit up the skies of London or New York, such as an air burst like the one that woke Bill Clinton? A hint can be found in the distant wilds of Siberia.

ABOVE **THE TUNGUSKA METEOR AIR BURST IN 1908 FLATTENED UNPOPULATED SIBERIAN FOREST, BUT IF IT HAD ARRIVED FOUR HOURS LATER IT WOULD HAVE LEVELLED ST PETERSBURG.**

The Tunguska Meteor

On 30 June 1908, at approximately 7:17 a.m. local time, an almighty explosion rocked central Siberia. Earth tremors registered on seismographs across the continent, while on the Trans-Siberian Railway a train halted for fear of being derailed by an earthquake. Instead, the stricken passengers saw the dawn sky lit up by a meteor fireball half the size of a full moon. The epicentre of the blast lay 650 km (400 miles) to their south in the virtually uninhabited Tunguska River area. No-one was killed, but a reindeer herder later returned to his 1500-strong herd to find only charred bodies, and a pile of stones where his hut had been. The few locals unlucky enough to be caught within a few tens of kilometres of the blast zone were bruised, battered, deafened and singed ... but alive. It wasn't until 20 years later that the political turmoil of revolutionary Russia had quelled sufficiently for the first scientific expedition to be allowed access to the area. What they found was 2000 square km (770 square miles) of Siberian forest – an area the size of Greater London within the M25 – laid waste by the explosion. Vast tracts of trees were flattened, their fallen trunks pointing inwards to a 'ground zero', where, remarkably, the trees had been stripped of their branches but were still upright, like a forest of telegraph poles. These clues helped the Russian scientists reconstruct what had happened. It seems that a meteor stream exploded 6–10 km (4–6 miles) above the

How Old is Earth?

Just as the ticking of a clock, second by second, sets the pace of our human lives, so the crackle of the Geiger counter provides the timepiece for our planet. A Geiger counter detects the radioactive decay of certain natural elements. These radioactive elements are nature's changelings, capable of existing in different forms, or isotopes, that transform their atomic make-up with unerring fidelity, shedding neutrons like clockwork. Understanding this atomic alchemy is no black art but instead emerged from the nuclear physics that also gave us atom bombs and nuclear power. What it gave geologists was, quite simply, the lifespan of our planet.

Different radioactive isotopes decay at different rates, varying in their half-life – the time it takes for half a given sample to disintegrate into another substance. Some isotopes undo themselves in a geological instant. Americium-241, the radioactive element used in smoke detectors, loses half of its atoms every 458 years, while the carbon-14 isotope used to date archaeological remains has a half-life of 5700 years. Others are remarkably enduring. The isotope uranium-238, which is present in tiny amounts in rock, has a half-life of 4.5 billion years. It decays to form lead-206, so if we measure the amount of both isotopes in a rock we can quickly calculate its age.

For much of the nineteenth century, Earth's age was pegged at around the 100 million year mark, thanks mainly to the work of Scottish physicist William Thomson (later Lord Kelvin), who based his calculations on the time it should take for Earth to cool down from a hot, molten ball. But in the first few years of the twentieth century came a sensational claim from New Zealand-born physicist Ernest Rutherford, who announced that a particular sample of the radioactive mineral pitchblende was 700 million years old, based on its rate of decay. It was apparently far older than Earth itself. Within a few years, the British geologist Arthur Holmes had used isotopes from the uranium–lead family to calculate the age of our planet's major rock strata, giving us the first geological timescale. The dates have since been refined, but Holmes was remarkably close to modern figures. He put the start of the Cambrian Period, for example, at 500 million years ago, whereas we now put it at 540 million.

The new science of nuclear physics gradually pushed the age of the oldest rocks further and further back. Even by 1910, British physicist Robert Strutt had showed from the decay of thorium that a mineral sample from Sri Lanka was more than 2400 million years old, making Earth, in terms of age, a multibillionaire. But it was a meteorite that gave the ultimate prize: the age of the planet itself. In 1948,

BELOW LEFT **GEOLOGIST JAMES HUTTON**
BELOW RIGHT **SCOTTISH PHYSICIST WILLIAM THOMSON (LATER LORD KELVIN)**

American geochemist Clair Patterson used uranium and lead isotopes to date fragments of the Barringer Crater meteorite, which, since it was a leftover from the early solar system, must have formed around the same time as Earth. The result was so surprising that he thought he would have a heart attack and considered checking himself into hospital: it was a mind-blowing 4.55 billion years. Give or take a few tens of millions, it's the age that still stands today.

Our solar system was now not only vast, it was also prodigiously ancient. For geologists, such antiquity meant that there had been enough time for our planet's geological features to have formed by the imperceptibly slow processes that happen all around us, settling an old dispute about whether Earth had developed gradually or through sudden catastrophes. But the sheer vast expanse of our planet's past is difficult to comprehend – after all, how do you imagine what seems like eternity? To the eighteenth-century geologist James Hutton, whose gradualist theories were vindicated by the discovery of Earth's antiquity, the indefinite realm of time conjured 'no vestige of a beginning – no prospect of an end'. The modern American writer John McPhee coined it differently, simply calling the immense abyss of Earth's past 'deep time'.

In his book *Time's Arrow, Time's Cycle*, the esteemed American palaeontologist Stephen Jay Gould offered perhaps the most resonant of metaphors, compressing 4.5 billion years of planetary history into a 24-hour day. Our planet's birth takes place on the stroke after midnight, and the 'Cambrian explosion' – in which complex animals first start crawling about – doesn't happen until 10 p.m. Dinosaurs don't show up until after 11 p.m. and are snuffed out 20 minutes before midnight, while modern humans arrive on the scene in the last two seconds of the day. Human civilization – some 6000 years of empire, art, religion and politics – is squeezed into the last tenth of a second.

McPhee conjured up another version of the metaphor. Consider Earth's lifespan as an old English yard: the distance from the king's nose to the tip of his outstretched hand. One stroke of a nail file on the middle finger erases human history. But it was the writer Mark Twain who perhaps captured it best when, still in the first decade of the twentieth century, he wrote presciently that if the Eiffel Tower, then the sparkling new icon of the modern era, represented the world's age 'then the skin of paint on the pinnacle-knob at the summit would represent Man's share of that age; and anyone would see that that skin was what the tower was built for'.

BELOW LEFT **AMERICAN GEOCHEMIST CLAIR CAMERON PATTERSON** BELOW RIGHT **BRITISH GEOLOGIST ARTHUR HOLMES**

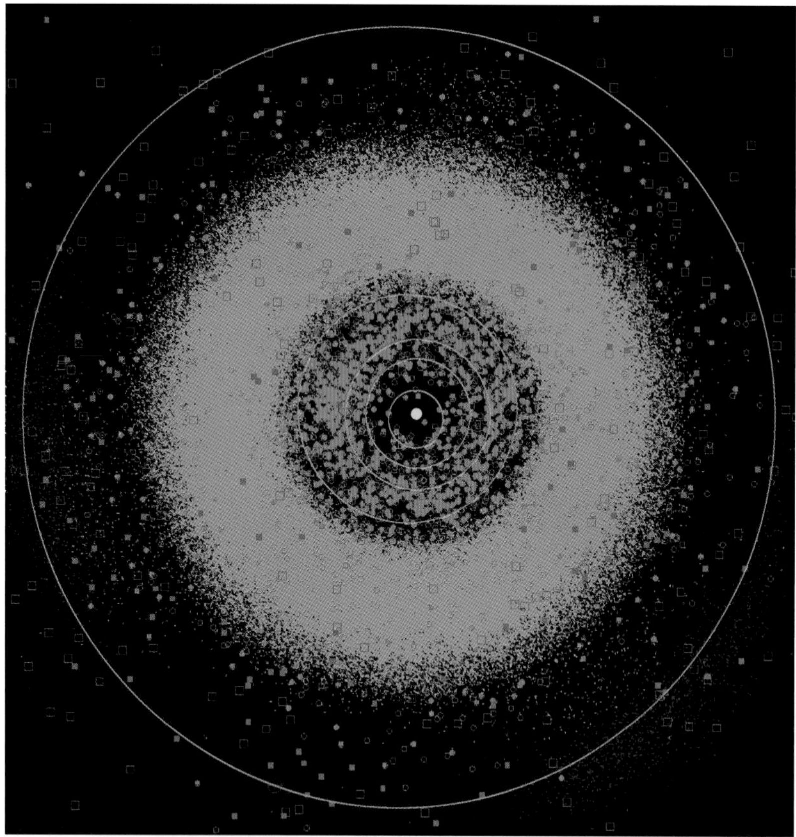

ABOVE **EARTH'S COSMIC NEIGHBOURHOOD IS BRIMMING WITH KILOMETRE-SIZED 'NEAR-EARTH OBJECTS' THAT STRAY CLOSE TO OUR PLANET'S ORBIT.**

ground, unleashing as much energy as a million tonnes of TNT or about 60 Hiroshima atomic bombs. And all this from something that probably measured barely 40 metres (130 ft) across – the size of a small block of flats.

If the Tunguska meteor had arrived 4 hours and 47 minutes later, Earth's rotation would have placed St Petersburg in its cross hairs, undoubtedly killing many thousands of people and perhaps snuffing out the city's bristling Bolshevik discontent. Twentieth-century history might well have been rewritten. Instead, what happened that summer's day in remote Siberia became no more than a historical footnote, and for the next few decades the world concentrated on home-grown devastation. In fact, it wasn't until the 1990s that scientists, in part galvanized by Shoemaker-Levy 9's assault on Jupiter, began to take seriously the threat of attack from above. In the last decade or so there has been a concerted international effort to catalogue asteroids and comets in orbits that pose a threat to our planet. The hunt has led to the realization that our cosmic neighbourhood is positively brimming with extraterrestrial missiles. For example, we now know that around 50 Tunguska-sized objects pass between Earth and the Moon every day, and we should expect a Tunguska-sized strike on average once every 50–100 years. Every 6000 years – roughly the length of time that has passed since civilization began – there should be a handful of far larger collisions, caused by objects hundreds of metres across. The effects of such megablasts would be far more dramatic than Tunguska: millions of tonnes of meteorite fallout would envelop the globe in a cooling dust veil, causing crop failures, famine and disease. It would be like a nuclear winter.

Some scientists believe that such a 'cosmic winter' happened during the remarkable decades of the mid-sixth century, appropriately enough during Europe's Dark Ages. Tree rings in oak timbers from all over northern Europe show virtually no growth in the years 535–545 AD, suggesting there was a very severe chill, and the summer of 536 appears to have been the second coldest in the 1500-year tree-ring record. It was during this period that the Justinian Plague wiped out much of Constantinople and the Black Death seems to have made its first appearance in Europe. Chinese chronicles record great falls of yellow dust towards the end of 536, and in the next few years there was famine and disease throughout China, Korea and Japan. Evidence for cooling stretches the length of the western Americas, from California to Chile, supporting the notion that this was a worldwide crisis. Some geologists think the sudden chill was caused by a giant volcanic eruption, but while volcanoes in Iceland, Indonesia and Papua New Guinea have all come under suspicion, no-one has yet found a convincing smoking gun. Meanwhile, astrophysicists argue that the frigid spell could easily have been produced by the fallout plume from the break-up of a comet 600 metres (2000 feet) wide.

Today, even with all our sophisticated telescopic monitoring of the night sky, a lump of rock measuring a few hundred metres across would probably remain undetected until too late. Although astronomers are actively hunting down 'near-Earth objects' (NEOs) that stray across the inner solar system, their efforts are focused solely on boulders more than 1 km (0.6 miles) wide. These giants have the potential to strike with a hundred times the explosive force of the world's combined nuclear arsenal, thereby terminating civilization as we know it. In 2000, up to 1000 such NEOs were believed to be lurking in our planet's backyard; by 2006, telescopic searches had found and tracked more than 700 of them. Only a handful had trajectories that gave any more than a slim chance of collision with Earth this century, so the likelihood of a civilization-ending event appears to be very small. However, for every lump of space debris larger than a kilometre, there are thousands that are hundreds of metres wide, and tens of thousands that are tens of metres wide – and these deadly objects aren't being tracked at all. Deep in the heart of the Arizona desert, there is dramatic proof of what these unseen missiles can do.

Impact Scars

In 1911, Daniel Moreau Barringer was a man consumed with a staggering idea: that a large hole in the desert scrub of eastern Arizona held an extraterrestrial bonanza. Iron meteorite fragments had been found scattered in a dry river bed called Diablo Canyon, close to an enigmatic, bowl-shaped crater measuring just over a kilometre in diameter. To the entrepreneurial engineer Barringer, this meant there must be an enormous and valuable lump of iron buried in the heart of the crater. But to the leading geologists of the day, Barringer's theory was baseless fantasy – the circular depression was probably the result of some kind of steam explosion. Undaunted by geological scoffing, and lured by the riches he thought lay under the crater floor, Barringer spent 26 years and $600,000 surveying and drilling the site, but to no avail. Shortly before his death in 1929, scientists worked out where his missing metal was: it had been vaporized by the collision. The fallen halo of metal fragments that had drawn Barringer to the crater was actually all that was left of it. The discovery was of little consolation to Barringer, who died virtually penniless. His ultimate legacy, however, would be the crater itself. Barringer Crater was named after him and is still held in the family name. It was the first scientifically accepted meteorite crater on Earth.

Today, Barringer's crater is better known as Meteor Crater, the major tourist draw for folks travelling along Route 66 through the former railroad stop of Winslow. The rock band the Eagles immortalized this staging post in their song 'Take it Easy', but it is the large hole in the ground to the west of the town that has achieved true fame. For a start, fragments of what became known as the Canyon Diablo meteorite would eventually enable geologists to figure out the true age of our planet (*see* 'How old is Earth?', page 50). But it is the crater's role in opening our eyes to the true nature of cosmic collisions that ensures the enduring fame of Barringer's folly. The tourists who wind their way up the crater rim to stand in awed silence doubtless try to visualize how, 50,000 years ago (I think it was a Tuesday), a lump of rock and metal just 30–50 metres (100–165 feet) wide managed to excavate a cavity 200

metres (650 feet) deep. Listen to the guides and they'll tell you the incoming projectile was travelling fast enough to get from Los Angeles to New York in four minutes – just long enough to boil an egg. The ballistics are truly mind-blowing, but our understanding of precisely what happened at the moment of impact comes not from meteorite craters but from decades of blowing stuff up.

Nuclear Intelligence

In the 1950s and 1960s, scientists who had witnessed first-hand the effects of bombs and barrages during World War II turned their attention to the destructive might of nuclear devices and high-tonnage TNT blasts. For reference, the atomic bomb at Hiroshima exploded with the energy of 15,000 tonnes of TNT, and the hydrogen bomb tested at Bikini Atoll had the explosive force of 10 million tonnes of TNT. The data gathered from many such blasts has enabled scientists to create computer models that tell us what happens when a giant meteorite smashes into Earth. Falling at a typical velocity of 20 km (12 miles) a second (faster in the case of comets), the projectile has phenomenal kinetic energy, and this is unleashed in a catastrophic explosion. Even before the impactor touches down, its fiery plummet through the atmosphere has brought it close to melting. Ahead of it, a compressed column of air slams into the ground, followed immediately by the projectile. The stupendous pressure and heat generated by impact cause both the projectile and the rock at ground zero to disintegrate and vaporize. The vaporized rock explodes from the site as a spray of plasma before raining back to Earth in a shower of molten fallout and forming characteristic cone-shaped droplets of glass called tektites. Shock waves tear through the ground, penetrating the planet

LEFT **BARRINGER'S CRATER, BETTER KNOWN AS METEOR CRATER: A HOLE IN THE ARIZONA DESERT THAT LED ONE MAN TO FINANCIAL RUIN AND ANOTHER TO FAME AS THE DISCOVERER OF EARTH'S FIRST SCIENTIFICALLY ACCEPTED IMPACT CRATER.**

like an earthquake, and a lateral air blast overturns the surrounding rock strata, dispersing larger clumps of rock into an apron of debris to form a crater rim and flinging finer material into space.

Barringer Crater became the first place where effects that had only previously been observed on a battlefield or weapons testing range were found etched into the landscape and written in the rocks. As a sign on the road to the site proudly announces, this hole was the prototype for the scientific study of all impact craters in the galaxy. And the person who made it famous was the man who decades later would co-discover the Shoemaker-Levy 9 comet: Eugene Shoemaker. Already an expert on the effects of large explosions, Shoemaker carried out a painstaking ballistic analysis of the Barringer Crater, doggedly mapping the terrain, unravelling the strangely contorted rock strata, and examining the pulverized rock under the microscope. Among his discoveries was a raised lip around the crater rim, which had swollen up after the impact; an enormous flap of rock that had been flipped over by the blast; rocks that had been intricately shattered by the impact; and crystals of 'shocked quartz' within desert rocks that had been subjected to sudden high pressure. What emerged from Shoemaker's ground-breaking forensic work was quite simply the complete anatomy of a terrestrial impact crater. And once the telltale symptoms of an impact were understood, meteor craters began turning up all over the place.

At the most recent count, there are 171 authenticated impact craters on Earth, the vast majority of these being on land. The small ones – less than 4 km (2.5 miles) in size – share Barringer Crater's simple form: a circular bowl lightly dusted but not buried by fallout debris. Ironically, the circular shape long convinced many geologists that such craters could not have been excavated by asteroids or comets. As the impact sceptics pointed out, meteors generally arrive on inclined trajectories and so ought to make elliptical craters. But Shoemaker demonstrated that small impactors travelling at high speeds vaporize explosively, producing a blast that radiates equally in all directions and so carves out a circular hole. For craters wider than 4 km (2.5 miles), studies of nuclear explosions have shown that a more complex architecture can be expected. For a start, large craters often have an uplifted centre surrounded by a trough, as well as a broken rim. The uplifted centre forms because the ground rebounds after impact. It is something like the ripple pattern that forms when a drop of water splashes into a pool, though when molten rock is involved the pattern becomes frozen as the rock cools and solidifies. Colossal examples of such complex craters adorn the Moon, and equivalent features have been found here on Earth. The largest of them are the 300 km (190 mile) wide Vredefort Crater in South Africa and the 250 km (155 mile) wide Sudbury Basin in Ontario. Unlike Barringer, these are truly ancient scars, relics of collisions that happened around 2 billion years ago.

OPPOSITE **A SELECTION OF EARTH'S IMPACT CRATERS. CLOCKWISE FROM THE TOP: GOSSES BLUFF, NORTHERN TERRITORY, AUSTRALIA; SHOEMAKER (FORMERLY TEAGUE) CRATER, AUSTRALIA; AOROUNGA IMPACT CRATER, CHAD.**

Impacts at Sea

Given that water covers three-quarters of our planet's surface, it might seem surprising that only 10 per cent of Earth's known impact craters have been found in the ocean. The trouble is that marine impact scars are difficult to find, for two main reasons. Firstly, Earth's ocean floors are surprisingly young, having mostly formed within the last 200 million years, so only recent impacts would be preserved. Secondly, craters on the ocean floor look different from those on land and quickly get buried by sediment. Nevertheless, the ever-expanding search for oil and minerals in the ocean depths has inadvertently uncovered evidence of some dramatic impacts. And with this comes a growing appreciation of the lethal threat these watery landings might bring.

Impacts at sea are different from those on land in one major respect: they create waves. In fact, they create giant tsunamis. For objects larger than a kilometre wide, the water slows but doesn't stop the missile, which continues all the way to the sea bed. The entire depth of the water column is displaced by the impact, instantly creating a wave whose height is equal to the ocean's depth at that point. In most cases that means a wave about 3 km (2 miles) tall, though it could be even bigger. This giant wall of water quickly collapses as the wave radiates away from the splash site, but the tsunami remains monstrous. Since we've never experienced a marine impact of this nature, only educated guesswork can tell us how much destruction it might cause. One computer simulation of a 5 km (3 mile) wide rock hitting the middle of the Atlantic Ocean predicts a monster wave swamping the eastern US all the way to the Appalachian Mountains and drowning Florida almost entirely. To the east, surges hundreds of metres tall overrun the coasts of Portugal, southern Spain and western France, and the wave enters the Strait of Gibraltar as a huge hydraulic bore before dissipating along the Mediterranean shores of Spain and Morocco. Northern Europe is not as badly flooded as Iberia, mainly because its wide continental shelf bears the brunt of the tsunami's energy, but it would be a matter of indifference to someone in Cork, Ireland, to be drowned by a mere 20 metre (65 foot) wave instead of the towering wall of water engulfing Lisbon or Cadiz. And the wave isn't finished yet: spreading out from the Atlantic, it lashes virtually every stretch of coast in the world. That's the thing about ocean impacts – the reach of their destruction is truly global. And nothing shows this better than the extraordinary impact that happened at Chicxulub on the coast of Mexico's Yucatán peninsula 65 million years ago.

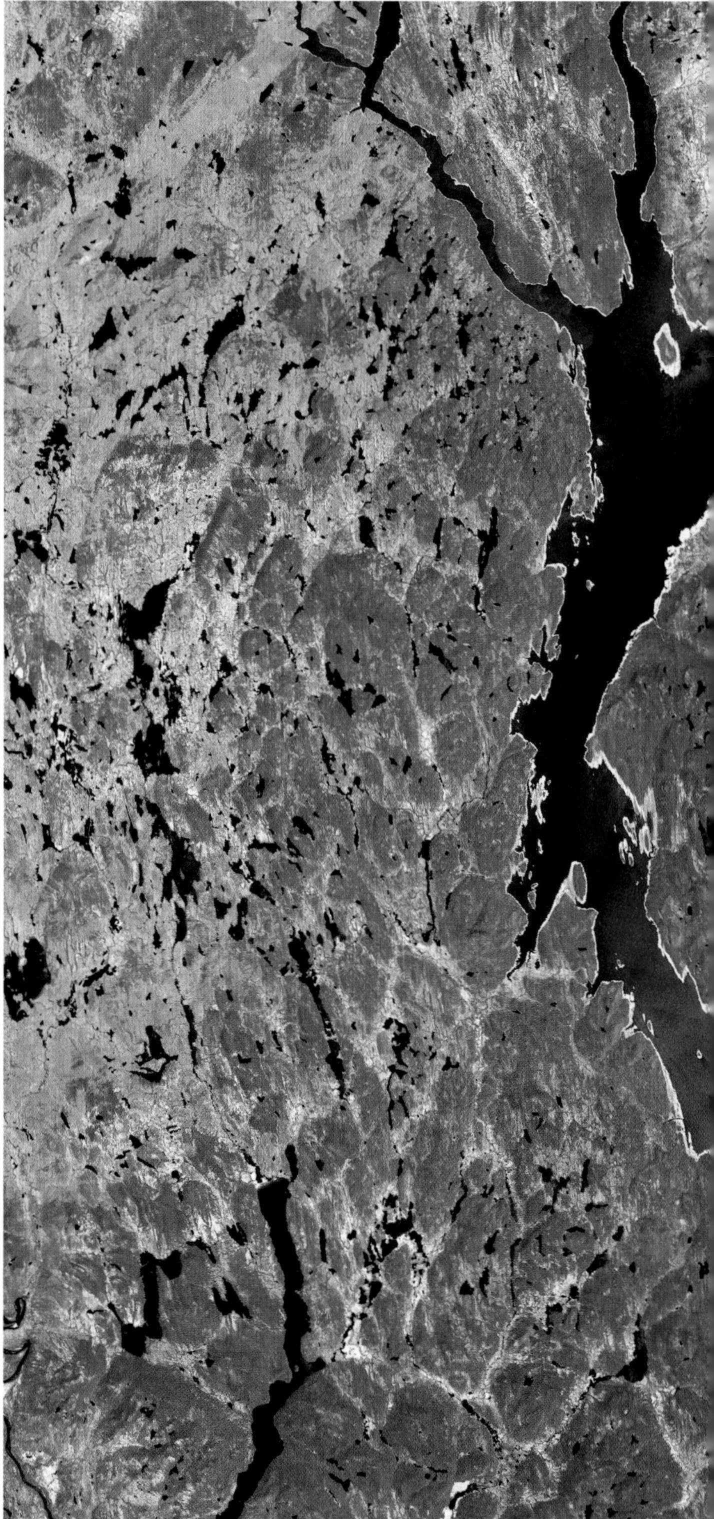

RIGHT **THE MANICOUAGAN CRATER IN QUEBEC IS ONE OF THE WORLD'S LARGEST AND OLDEST KNOWN IMPACT CRATERS.**

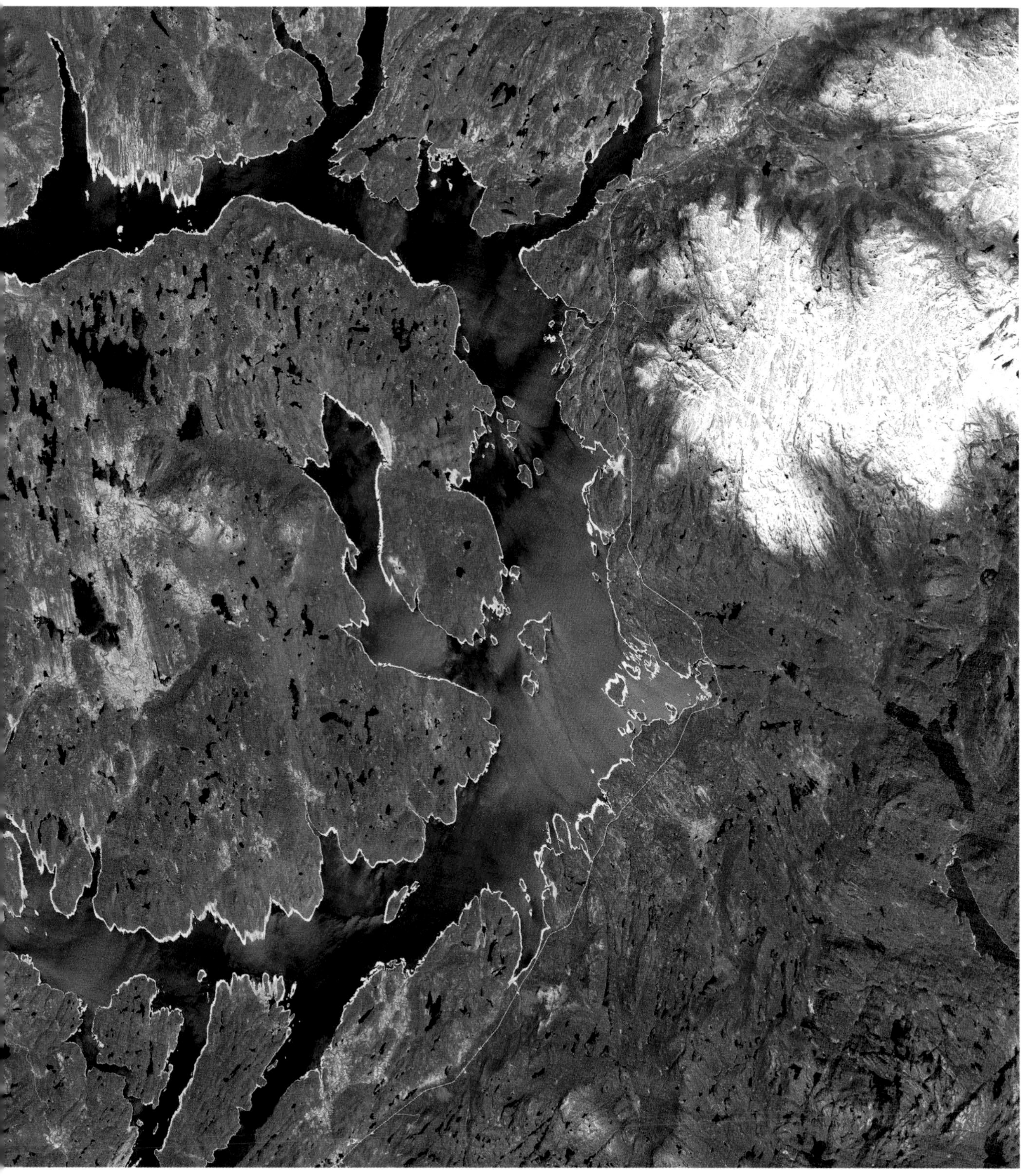

ABOVE **AN IMPRESSION OF HOW AN EARTH IMPACT EVENT MIGHT HAVE APPEARED.**

The End of the Dinosaurs

The asteroid heading for Mexico wouldn't have been visible from Earth until it hit the edge of the atmosphere and began to heat up. Travelling at cosmic velocity, it struck the surface a second later, compressing the air beneath it with such violence that the temperature soared to 6000°C (11,000°F) – as hot as the surface of the Sun. Estimated to be about 10 km (6 miles) wide – about as big as Glasgow – the rock struck with the energy of 100 million million tonnes of TNT, instantly vaporizing itself. You would need to detonate a Hiroshima-sized bomb for every person on Earth in order to generate a comparable explosion. Every living thing within 250 km (155 miles) that was not boiled alive would have been killed by the blast. From the epicentre came a shock wave that moved at almost the speed of light, sweeping everything out of its way and killing every plant and animal, the latter by causing haemorrhaging and oedema in the lungs. Forests were flattened for up to 2000 km (1200 miles) around, from the highlands of central Mexico to the Gulf states of the United States. Nearly every living thing within 1600 km (1000 miles) would have been either knocked down or ablaze. Enormous tsunamis radiated across the ocean, sending waves that were hundreds of metres tall deep into Texas and dragging the wreckage of forests back out to sea. The whole region was ravaged by great earthquakes and 'hypercanes' – super-hurricanes with 800 km/h (500 mph) winds driven by the energy of seas that had been heated to 120°C (250°F). After the blinding flash of impact, a veil of darkness raced silently across the sky, moving faster than the speed of sound, and a thick blanket of debris buried everything within a few hundred kilometres of ground zero.

The blast excavated a cavity 100 km (62 miles) in diameter, blowing out 1000 cubic km (240 cubic miles) of sulphur-rich rock, earth and superheated gas. Within an hour, the spreading cloud of debris had enveloped the globe, and burning rock debris was pelting down and setting much of the planet on fire. Ancient air bubbles trapped in amber fossils from this period show that the level of oxygen in the atmosphere was much higher than today, so the fires would have been all the more ferocious, darkening the skies with smoke. Charcoal and soot deposits around the world testify to the scale of the inferno, which must have stripped vast swathes of forest from the land, consuming animals by the million and robbing many more of their food. Smoke and impact debris would have blotted out the Sun for months, killing off plant life and disrupting food chains on land and in the oceans. In the months that followed the impact, an estimated 600 billion tonnes of sulphuric acid fell as acid rain caustic enough to burn skin and defoliate plants. Equally noxious nitric acid rain asphyxiated animals and poisoned lakes and shallow seas. The cold and dark cosmic winter probably lasted only a matter of years, but the release of large quantities of greenhouse gases from vaporized sediments may have produced a long-term rise in global temperatures. To add to the environmental mayhem, huge emissions of chlorine gas from vaporized rocks would have eaten away at Earth's protective ozone layer.

Biological Apocalypse

After the impact came the catastrophe. The biological consequences were truly apocalyptic: entire ecosystems collapsed, setting off a chain reaction that drove perhaps 65 per cent of the world's species to extinction. In the oceans, thick-shelled bivalves, reef-building corals, ammonites and many types of plankton disappeared simultaneously. Not all sea dwellers fared so poorly. Invertebrates living in the shallows, creatures dwelling in the blackness of the ocean depths, and organisms that relied on detritus for nourishment survived in sufficient numbers to recover quickly. On land, the ferns, so adept at surviving wildfires, quickly colonized fire-torn regions and re-established some vegetation cover. But the most famous land dwellers affected by the impact were of course the dinosaurs. They had risen to prominence on the back of an earlier impact that reset the evolutionary button 251 million years ago. This time they were the victims, and our mammalian ancestors were the beneficiaries.

Or that's the story. In recent years, things have become increasingly tangled. Since the discovery of the Chicxulub Crater in Mexico, other craters have been found to date from the same time, such as the Boltysh Crater in the Ukraine, the Eagle Butte Crater in Canada, the Silverpit Crater in the North Sea, and perhaps even the huge Shiva Crater in the Indian Ocean. A cluster of collisions had seemed highly unlikely until Shoemaker-Levy 9 was seen delivering blow after blow to Jupiter. Had Earth suffered a similar fate?

And yet there was another possibility, one that involved a different way of looking at the problem. What if there was a way to create worldwide devastation without recourse to the heavens? After all, there were at least four mass extinctions of life prior to the one in which the dinosaurs perished, and none of these has been convincingly linked to an impact. In fact, some of Earth's largest impact craters appear to have formed with very little ecological disturbance. Given the number of craters on Earth and the frequency with which large impacts might be expected to occur, life on our planet appears to have been remarkably untroubled by extraterrestrial strikes. These have undoubtedly occurred, but perhaps they do not exert the controlling influence we had suspected. Instead, there is another phenomenon that shows an almost perfect correlation with mass extinctions and is equally capable of wreaking global havoc. At the same time as the asteroid was crash-landing in Mexico 65 million years ago, vast outpourings of lava were spewing from fissures on the opposite side of the globe. The vast Deccan lava fields of India are a potent expression of a force deep inside Earth that is responsible for very many of our planet's distinctive features – a force that is the heat engine that drives our planet.

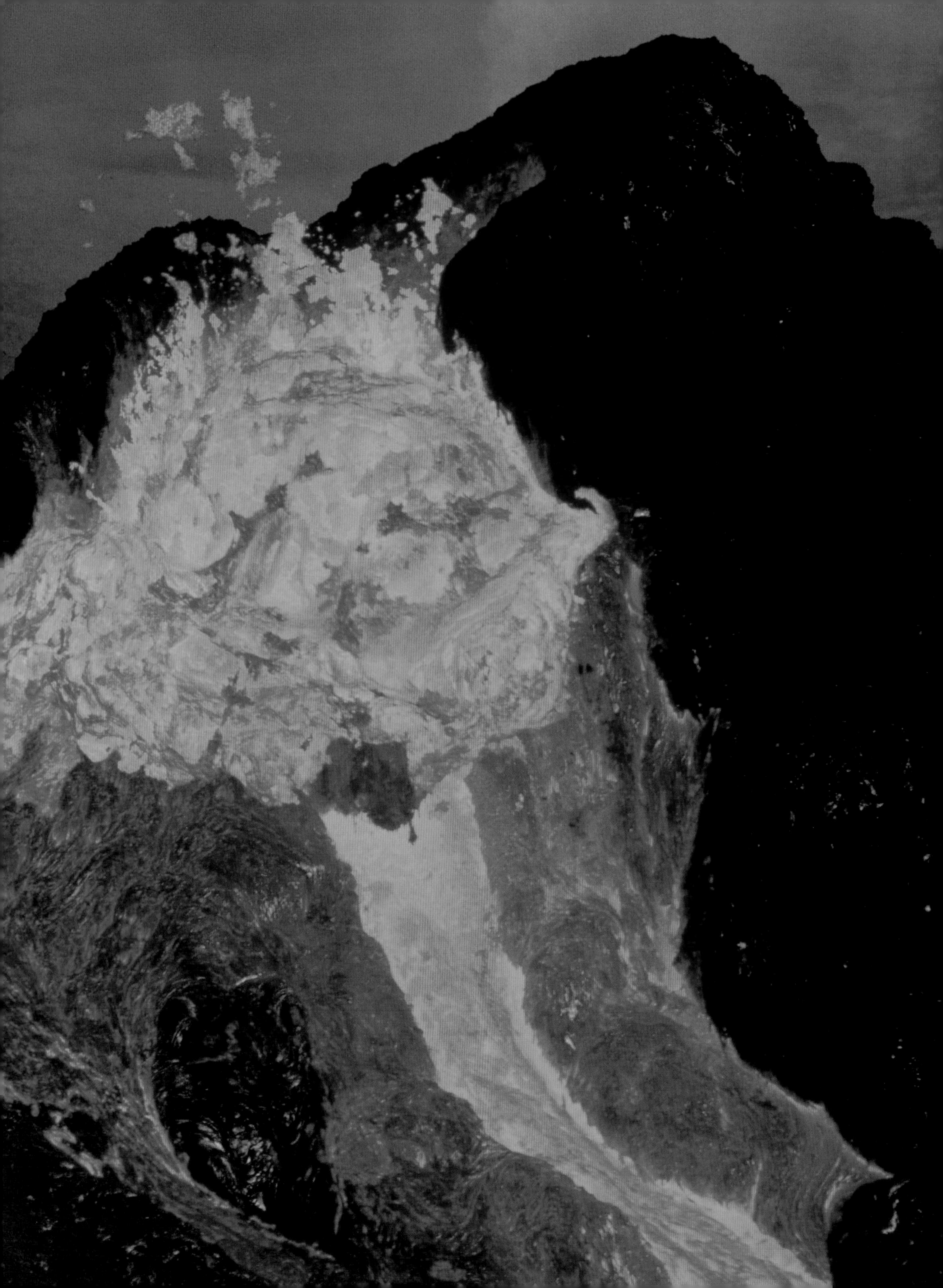

CHAPTER TWO

HEAT

ERTA ALE IS NO ORDINARY VOLCANO. SITTING IN THE MIDDLE OF THE BAKING desert of Ethiopia's Afar Depression – one of the lowest and hottest places on the planet – a permanent pool of magma bubbles out from Earth's interior. The lava lake of Erta Ale is among the least visited volcanoes in the world, one end of a gruelling three-day truck drive and a one-day camel trek, the last stretch through habitual bandit country. Only the perennial fires of Mount Erebus in Antarctica are more remote, and the active lava craters of war-torn Congo more dangerous to get to. In fact, the only secure way to reach Erta Ale is to hitch a lift with an Ethiopian military helicopter, though in this spluttering conflict zone on the Eritrean border they are often otherwise engaged. The flight is spectacular, sweeping first over the verdant fields of the Ethiopian highlands, then over steep canyons, which grade swiftly into the parched, sandy plains and dusty lake beds of the Afar Depression. It is astounding to think that this baking, impoverished land is one of the cradles of humanity, home to the world's oldest stone tools and the bones of some of our earliest human ancestors. Eventually, inhospitable desert gives way to a volcanic wonderland. Dark tongues of hardened lava trailing out from conical volcanic hills are a sign that we're in the middle of Africa's Great Rift Valley, where, for the last few million years, the continent has been tearing apart.

OPPOSITE **THE SEETHING MOLTEN WORLD OF THE ACTIVE VOLCANO, SUCH AS HERE IN HAWAII, IS THE CLOSEST WE CAN GET TO WITNESSING THE HIDDEN HEAT ENGINE THAT POWERS THE PLANET.**

Among the chain of volcanoes that have arisen here is the vast shield of Erta Ale, visible on the horizon as a low bulge only 600 metres (2000 feet) tall but 50 km (30 miles) wide. From closer, Erta Ale's two fiery summit craters come into view, fumes spewing out of them. In the local Afar language, Erta Ale means 'smoking mountain', and the cloak of gas and smoke makes the area look like a war zone. It's as if we're entering enemy territory.

The ground, which seemed silky smooth from the air, is a twisted and tortured carapace of solidified lava. The first few stumbles on the sharp-edged rock draw blood. Much of the surface underfoot is covered in loose and broken 'aa', a Hawaiian term for the rough clinker that often forms the top of moving lava flows. Yet amidst the jagged minefield there are also wonderful sculpted swirls of pahoehoe, the smooth, undulating and occasionally ropey rock formed from very fluid lava. Picking carefully over the jumbled orgy of frozen magma, it is a walk of only a few hundred metres to the rim of the large northern crater, with some gaping wrenches and crevasses to be gingerly crossed near the edge. From the precarious lip of a crumbling cliff 30 metres (100 feet) tall, you peer down into an enormous bubbling, hissing pit of lava.

By day, the harsh sunlight makes the bright lava hard to see clearly, not least because it is often clouded over by the gases that billow within the crater and spill over its jaws. Water vapour, carbon dioxide and sulphur dioxide are the main ingredients in the noxious cocktail of fumes being expelled here. The vapours come and go in warm, choking gusts, almost like breaths, and a gas mask is *de rigueur*. At night, amidst the eerie red-orange glow of the crater, gas masks are kept to hand because the fumes can waft down the slope and creep into tents. Living for days in this acrid scene, it is easy to imagine you've been transported back in time, back to the very earliest days of our planet.

RIGHT **THE MAGMA LAKE AT REMOTE ERTA ALE VOLCANO IN ETHIOPIA – ONE OF THE MOST SPECTACULAR IN THE WORLD.**

Earth Heats Up

To understand where our planet's inner heat comes from, we have to go back 4.5 billion years to the time of Earth's birth. Geologists call Earth's early years the Hadean Era after the Greek word for hell. And for good reason, because the newborn Earth was a cruel and hostile world, a land of fire and brimstone. But while the primordial world is widely and enduringly depicted as a land of exuberant volcanoes, with red-hot oceans of magma and sulphurous skies lit up by meteorite fireballs, it may not have been the fiery Hades of popular imagination. Granted, asteroids and comets pummelled Earth from space, delivering massive bolts of energy, but only a small proportion of the heat they released would have lingered at the planet's surface. In fact, between impacts the young Earth was mostly a Norse Ice-Hades, a frigid world only occasionally interrupted by intervals of true inferno. Deep inside, however, Earth was warming up.

Stoked by profligate bursts of heat from extraterrestrial bumps and bangs, and by thermal energy released from radioactive elements, temperatures rose far faster inside the young Earth than they could be lowered by slow conduction. Within about 10 million years of the planet's birth, the average temperature of the interior had soared to some 2000°C (3600°F) – hot enough to melt virtually the entire planet. In a largely molten planet, gravity will separate denser elements from lighter ones. One of the densest elements in Earth's interior was iron, which was also by far the most abundant element present. Being heavier than the surrounding matter, the molten iron formed droplets that sank slowly towards the planet's centre. As they sank, gravitational energy was converted into heat energy, putting fire in the belly of the new planet. Gradually, Earth developed its heat engine.

In the first few hundred million years of its existence, Earth changed from being a hellish place to live to being just a lousy one. Above ground, as we'll discover in later chapters, oceans and atmosphere gradually gave the planet a complete makeover. Below ground, equally dramatic changes had taken place. Like a giant glass of Guinness separating into black stout at the bottom and white froth on top, Earth had separated out to form an inner metallic core and an outer rocky mantle. And stoked by its hot, metallic heart, Earth now had a circulatory system to carry heat around its body.

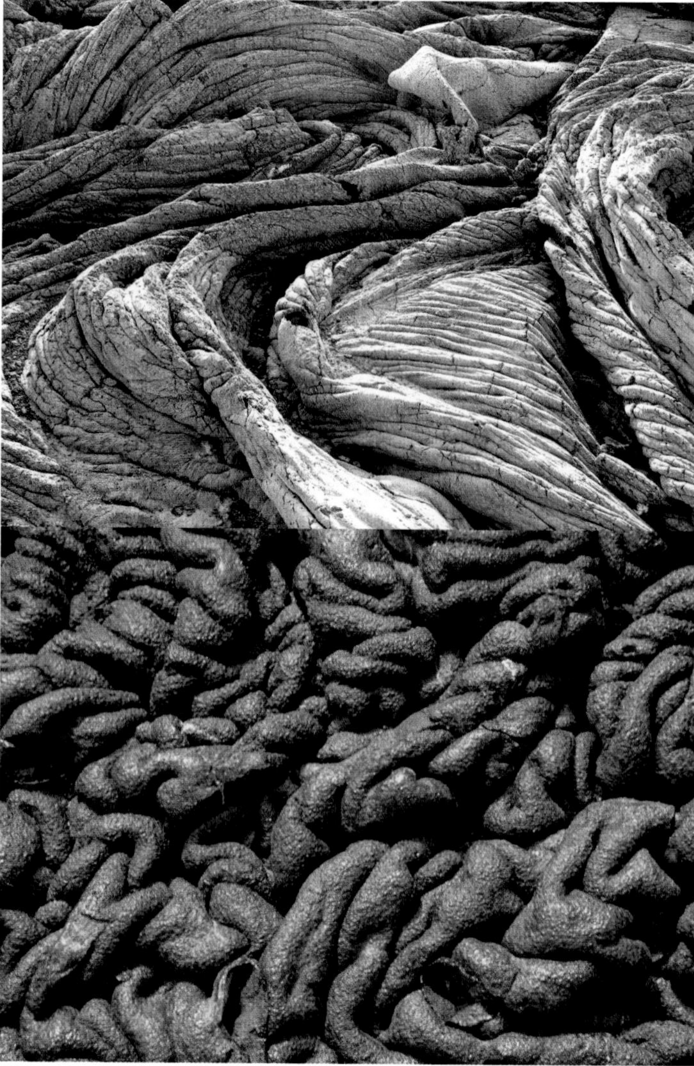

ABOVE **THE PETRIFIED TWISTS, WHORLS AND FOLDS OF ONCE FLOWING LAVA ALLOW VOLCANO SCIENTISTS TO RECONSTRUCT THE DYNAMICS OF PAST ERUPTIONS.**

In any body of liquid that is hotter at the bottom than the top, the hotter material expands and floats upwards over denser, cooler material until, in turn, it cools, becomes more dense and sinks again. This simple thermal yo-yo, called convection, is the mechanism that drives Earth's heat engine.

With its temperature regulated by the ups-and-downs of convecting currents, Earth soon cooled enough for its outer regions to solidify. So too did the planet's innermost core; over hundreds of millions of years, the central part of the core was squeezed into a solid lump of metal by pressures 3.5 million times greater than those at the surface. Temperatures down in that engine room, roughly 6000 km (4000 miles) beneath our feet, may be as high as on the glowing surface of the Sun: an incredible 5000–6000°C (9000–11,000°F). The energy is mostly just leftover heat that has lingered since the planet's formation, though there is also a modest heating from natural radioactivity. The primordial heat of the core is the central heating system for our planetary home.

Now you may think that our planet's innermost recesses are a long way from your everyday concerns. But you'd be very wrong, because it turns out that Earth's scalding core is our first line of defence against attack from space. The core did not entirely solidify after it formed – with lower pressures away from the centre, the outer part remained molten. Which is just as well, because our molten outer core creates the magnetic field that wards off lethal cosmic radiation (*see* 'A Magnetic Attraction', page 68).

Journey to the Centre of the Earth

The seething molten world of the active volcano is the closest we can get to witnessing the furnace that burns deep inside Earth. Standing on the rim of Erta Ale is like looking into the heart of our planet, straight down to its very core. But to really understand what is going on and to see how convection works, you need to get even closer to the magma.

The lava lake at Erta Ale has been continuously active for the last century, but the level of the magma rises and falls over time, sometimes retreating down the subterranean well that feeds the lake, at other times surging up to overtop the walls. The top of the crater rim formed 30 years ago in an enormous outpouring that resurfaced virtually the entire summit, while 30 metres (100 feet) or so below the rim there is a wide ledge of solidified lava that formed when the lake flooded over in 2004. The surface of the lake is a few

A Magnetic Attraction

Twice a year, usually around February and October, the long, dark nights of the Arctic winter are set ablaze by a spectacular neon light show. The famous northern lights, or aurora borealis, are best seen from the high-latitude 'auroral ring' that passes through Siberia, Alaska and northern Scandinavia, but they are not uncommon at lower latitudes and are occasionally seen above the equator. A twin phenomenon, aurora australis, illuminates southern skies inside the Antarctic Circle. In fact, although the light show is brightest near the poles, Earth's auroras can envelop the whole planet. And that's because they are caused by a titanic clash between the energy of Earth's molten interior and the power of the Sun.

ABOVE **THE AURORA AUSTRALIS SEEN FROM SPACE.** OPPOSITE **THE AURORA BOREALIS SEEN FROM CANADA.**

The Sun's stormy surface regularly flings out huge gusts of solar particles, or plasma. Hurled across space at up to 1000 km (620 miles) per second, they take a mere 2–3 days to reach our planet. About 60,000 km (37,000 miles) from Earth, the stream of charged particles meets the outer reaches of Earth's magnetic field, which deflects the plasma towards the magnetic poles. Closing in, the solar particles smash into Earth's atmosphere, colliding with air molecules and surrendering their energy as photons of light. It takes about 100 million photons to create a visible aurora, which usually takes the form of a shimmering curtain of light that dances across the night sky. When the Sun's storms are exceptionally violent, enough solar particles breach Earth's magnetic shield to light up temperate and even tropical skies too.

Sun storms can play havoc with electric systems on Earth. In March 1989, for instance, a magnetic disturbance cut off electricity supplies to 6 million people in Quebec and put a nuclear power station out of action for almost two days, while, high above, four US navigation satellites had to be taken out of service. But without Earth's magnetic field to protect us, things would be much worse. We would be continually bombarded with so much lethal cosmic radiation that we'd have to wear lead suits or live in caves.

It's tempting to imagine that Earth's magnetic field is generated by the planet's iron core in much the same way as a bar magnet works, but iron loses its magnetism when it heats up, and the core is far too hot. Instead, it is thought to be the swirling motion of the liquid outer core, stirred by Earth's rotation, that generates a magnetic field. The idea is that, very early in our planet's history, the slow-moving currents in the core passed through invisible lines of force in the Sun's magnetic field, causing electrons to twitch and move. This generated an electric current, which, in turn, induced a local magnetic field around Earth. Once stimulated, this home-grown dynamo was self-sustaining, with the mechanical energy of the flowing liquid now interacting with its own magnetic force lines to continually produce electromagnetic energy. Given the forces involved, you might think Earth's magnetic field is remarkably powerful, but in fact the opposite is true. Even a toy horseshoe magnet generates a magnetic force several hundred times more powerful than the field we live in, and the instruments needed to measure Earth's magnetism are so sensitive that a ticking wristwatch upsets them.

Earth's magnetic field is shaped as though the planet contains a gigantic bar magnet tilted about 11 degrees from the axis of rotation. The magnetic poles are not fixed in place but tend to wander around over the course of a year. Every half a million years or so, they do something far more dramatic: they swap places, magnetic north becoming magnetic south. These reversals in the field must be due to changes to the fluid flow in the outer core, but what these changes are, and how quickly the whole field turns upside down, is not clear. There are some indications that the flip could happen in a matter of decades, but the truth is we just don't know. If it happened in our lifetime, we would probably see the strength of the magnetic north pole slowly waning as magnetic eddies developed in different parts of the world, before the force gathered strength again, this time in the south. The bad side of a magnetic flip is that we'd temporarily lose the protective shield that keeps solar storms at bay; the good side is that we might see exceptionally intricate and beautiful auroras all over the world.

metres below the ledge, making it possible to descend to the rock platform and walk right to the edge of the active lake.

A nervous, roped descent into the crater reminds one of the lengths that geologists often go to in order to reveal the inner workings of our planet. Today, volcanoes can be monitored by all sorts of high-tech imaging from space, while on the ground a suite of precision instruments can remotely measure temperatures, gas output, ground movements and the like. And yet grabbing a sample of lava rock for laboratory analysis or scooping out a blob of semi-molten magma for an on-site temperature probe is still the bread-and-butter work of modern volcano science.

More is known about the workings of outer space than about what lies deep beneath our feet. Space probes have ventured to the limits of our solar system (*Voyager 1* has currently travelled 80 times further than the distance between Earth and the Sun), and in 1996 the *Galileo* probe successfully dived 600 metres (2000 feet) into Jupiter's outer atmosphere. But the deepest anyone has ever drilled into Earth is just 12 km (7.5 miles) on the Kola Peninsula in Russia – only 0.2 per cent of the distance to the centre. No wonder, then, that fantastical speculations abound about what exactly is down there. In his book *At the Earth's Core*, the author Edgar Rice Burroughs, best known as the creator of Tarzan, conceived that the crust was only 800 km (500 miles) thick and surrounded a vast, hollow interior, accessible via an opening near the North Pole, in which dinosaurs and huge mammals roamed. A similar subterranean world of caverns, lakes and creatures was charted by Professor Lindenbrock in Jules Verne's classic *Journey to the Centre of the Earth*. Lindenbrock and his nephew Axel entered Earth's interior via the mouth of the Icelandic volcano Snaefell and exited courtesy of an explosive eruption at Stromboli volcano in Italy. Even today the lack of first-hand observations of our planet's interior means that 'hollow Earth' theories litter the Internet.

Only a fraction of the vast sums of money spent exploring the largely empty and unfathomably vast territories of the Cosmos are used to probe our planet's interior, even though down below is likely to be crammed with interesting stuff. Stung by the disparity, a leading planetary scientist recently made an audacious proposal. With the loose change from the funds required to launch a space mission, a million tonnes of liquid iron could be poured into an artificial fracture on Earth's surface. The density of the iron would create a self-sustaining crack that would slowly but inexorably extend down towards the centre of Earth, carrying a grapefruit-sized instrument probe that could send first-hand information back from the bowels of the planet. OK, the proposal was largely tongue-in-cheek, but if such a mission to Earth's core ever got under way, what exactly would we expect it to find?

The World Within

The inside of planet Earth can be envisaged in various ways. Some geologists liken it to a peach, with its hard stone as a core, surrounded by a fleshy mantle and covered in a thin skin, the crust. For others, an avocado is more appropriate, in terms of the relative dimensions of core and mantle, but the notion is the same. Still others prefer to compare it to an onion, for however we probe the planet's depths we seem to detect more and more layers. In simple terms, there are just three layers: a metal core, a mantle of plastic rock and a brittle rocky crust. As we know, the core, which extends from a depth of 2900 km (1800 miles) to the planet's centre, settled out into two layers: a solid inner core enveloped by a liquid outer core. The surrounding mantle forms the main bulk of the planet and is mostly made of compounds combining two of the most abundant elements on Earth: silicon and oxygen. These so-called silicate rocks have also been separated by convection into layers of different densities. The lightest and most easily melted of the mantle's constituents became the crust, floating up as a kind of froth that cooled and solidified.

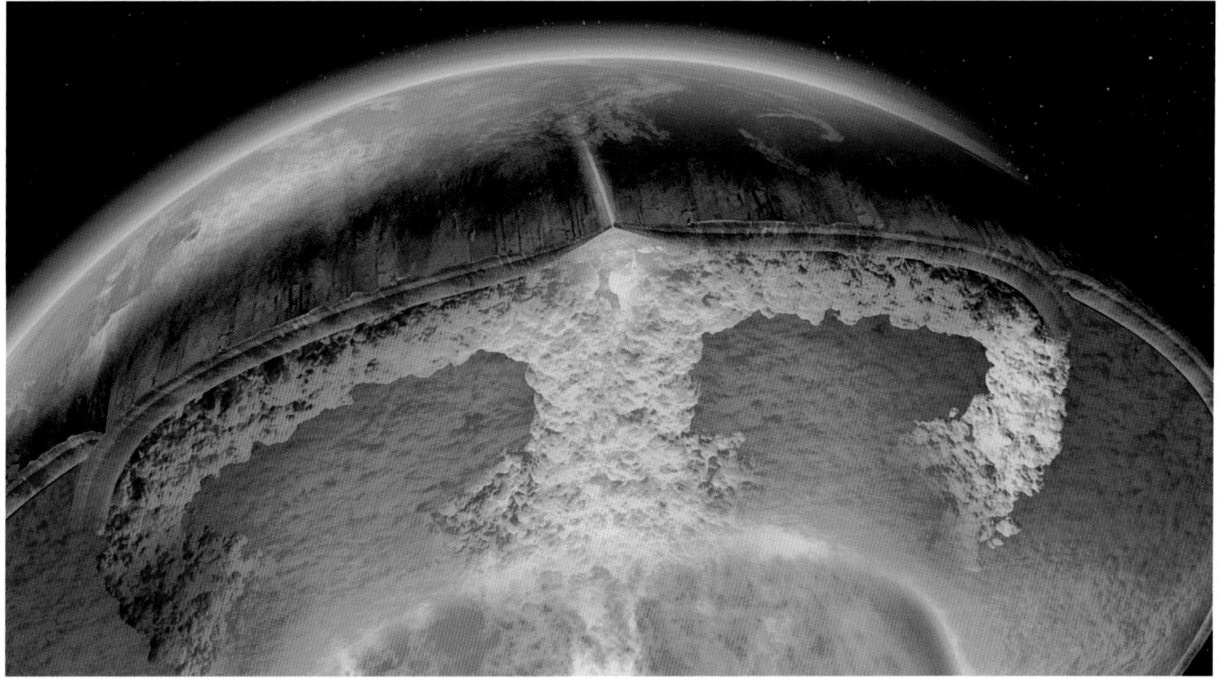

ABOVE **THE PERPETUAL CONVECTION OF HOT PLUMES OF ROCK FROM EARTH'S CORE TO ITS CRUST IS THE HEAT ENGINE THAT DRIVES OUR PLANET'S RESTLESS SURFACE.** OVERLEAF **RED-HOT FISSURES SPLIT THE DARK CRUST OF ERTA ALE'S LAVA LAKE – MINIATURE VERSIONS OF THE GREAT CRACKS IN EARTH'S RIGID SHELL THAT DEFINE OUR PLANET'S TECTONIC PLATES.**

Lightweight radioactive isotopes of potassium, uranium and thorium also rose with the crust, and these provide Earth's outermost layers with an additional source of heat. This shallow warmth explains why it quickly gets sweltering when you descend several kilometres into a deep gold mine; if the planet got hotter at the same rate all the way down, then its entire interior would be molten.

Think of Earth's mantle as a giant lava lamp heated by a light bulb, the core. In a lava lamp, heat from the light bulb in the base causes the waxy substance to expand, making it lighter than the viscous liquid above it and allowing it to rise. As the wax plume rises, it cools and slowly reverts to being more dense than the surrounding fluid, so it falls back down again, only to be reheated and for the cycle to be repeated endlessly. In just the same way, heat lost by the Earth's light bulb – the hot core – passes outwards to the mantle, warming the lowest layers and making them less dense. Hot, buoyant columns of strong but malleable rock, called mantle plumes, creep slowly upwards, rising about 2 metres (6.5 feet) a year, until they reach the base of the crust. Here, some of the heat and magma eventually escape through volcanic action, but most of the rock cools and descends again into the mantle abyss to await its next turn in the convective queue. It is this unrelenting rise and fall that transfers heat from the core to the surface and so forces the rigid outer crust to move.

World in Motion

More humans have walked on the Moon than have trodden on the fragile floor of Erta Ale's summit crater. The thin slabs of solidified lava, only two years old, crack easily underfoot, and close to the edge of the fiery pit there is the added danger of the fragile ground suddenly collapsing into the bubbling pool. But the danger is quickly forgotten when you see the extraordinary display

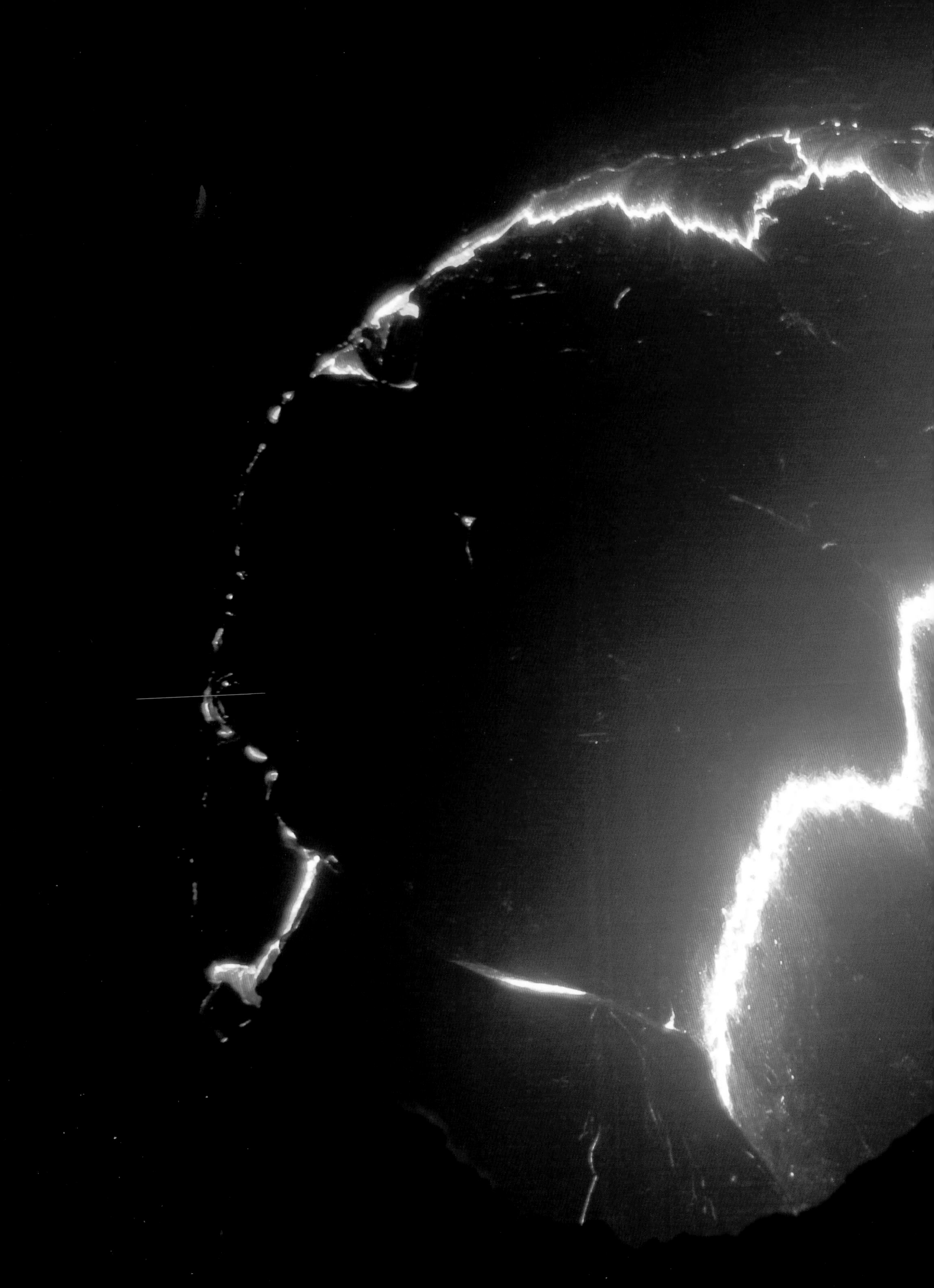

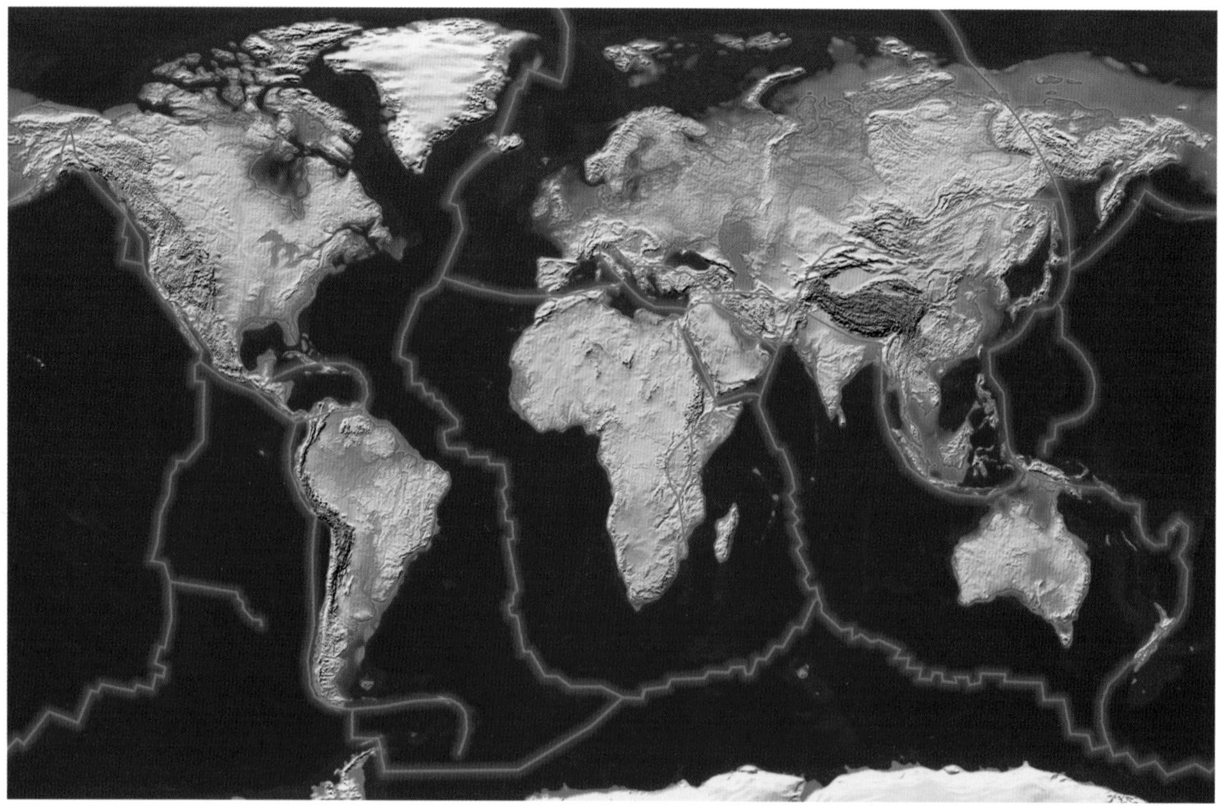

ABOVE **LIKE THE SEAMS OF A GIANT BASEBALL, THE SUTURES OF EARTH'S TECTONIC PLATES ENCIRCLE THE GLOBE. THESE LINES ARE THE MOST GEOLOGICALLY ACTIVE ON OUR PLANET – CORRIDORS OF EARTHQUAKES, VOLCANOES AND YOUNG MOUNTAIN RANGES.**

going on below. Every 15 minutes or so, a fiery ballet slowly unfolds on the lake surface. Initially, the surface has a thin crust of hardened, congealed lava, reddish black in colour, and with a temperature of about 350–450°C (660–840°F). It resembles a miniature version of Earth's rigid crust, complete with a network of tiny cracks, like crackles in the varnish of an old painting. Driven by material rising up, this crusted surface rips open, the dark craquelure riven by great fissures that spew forth bright orange magma from the still molten material below. The temperature of this molten rock is probably more than 1100°C (2000°F) – equivalent to the heat we would encounter in the great depths of Earth. Shouldered by convection currents, the upwelling lava carries a sheet of crust across the surface of the lake, until it founders and plunges near the lake's edge, instantly being remelted and consumed by the main body of the lava – reclaimed by the lake's 'mantle'. As the old skin disappears back down the hatch, a new fresh lava skin is congealing and thickening behind it. There is a temporary lull in the action. Then, suddenly, fresh incandescent cracks break the solid skin again, a giant boil of lava bursts out, and the dance is replayed, though always in a different manner. For all the blistering heat that the overturning lake is giving off, it is a hypnotic, entrancing performance.

What we are seeing played out in the turbulent rhythms of a lava lake barely 150 metres (500 feet) wide is the same pattern of relentless movement, the same cycle of destruction and renewal, that we see happening at the scale of whole oceans and continents on the face of the planet. This is plate tectonics. Stirred by convective upwellings of primordial heat from the deep, and stoked by added warmth from radioactivity in the crust, Earth's heat engine drives a moving mosaic of 20 or so rigid rafts, which are called plates. In some places, the plates just slide passively past each other. In others, particularly on the ocean floor, rising plumes blister a plate and it breaks apart; hot mantle rock seeps through the split and forms a new crust, widening the ocean floor and pushing continents apart. And if some parts of Earth's shell are spreading apart, others must be compressed together. When a plate of dense ocean floor is squeezed against a plate of buoyant continental rock, there is invariably only one winner. The dense, thin ocean floor generally gets pushed under the light continental mass and its rocks return to their mantle womb, carried to depths of a few hundred kilometres by a one-way escalator called a subduction zone. As the cold slab of oceanic rock sinks into the hot, plastic interior, dragging ever more ocean floor down with it, the whole ocean can get narrower until it eventually pulls closed. Then, with the sea gone, previously far-flung continents collide. Being equally buoyant, they crunch violently together, like a pair of wrestlers each unwilling to succumb to the other. The rocks become contorted, stacked and thickened into great mountain chains.

Plates move almost imperceptibly slowly, at rates of up to 12 cm (5 inches) a year. In a human lifetime we're talking just a few steps, but over the course of civilization it amounts to many tens of metres, and over millions of years the whole face of the planet becomes a restless congregation of jostling, shifting, separating and colliding parts. The motions seem to cycle from dispersing continental fragments around the globe to clustering them back together into a single landmass – a supercontinent – surrounded by a global ocean. Some 500 million years ago, far-flung continental landmasses were converging. One fleet of landmasses was drifting north from its moorings at the South Pole, bringing what would later become South America, Africa, Australia, India and Antarctica. In the northern hemisphere, two enormous landmasses were approaching each other, bringing together fragments that would later become Europe, Greenland and North America. The northern and southern landmasses then slowly began closing in on each other, shrinking the prehistoric ocean between to form a narrow tropical seaway. Finally, about 250 million years ago, the ocean sealed shut, and all the world's continents fused into a single continental amalgam called Pangaea – an enormous landmass that straddled the equator.

ABOVE **HOW EARTH MIGHT HAVE LOOKED 250 MILLION YEARS AGO, WITH ALL THE PLANET'S LANDMASSES FUSED INTO A SINGLE SUPERCONTINENT, PANGAEA.**

Pangaea was just one of perhaps six lost worlds that have come and gone with the ebb and flow of

ABOVE **A LAND BEING TORN APART AT THE SEAMS: ETHIOPIA'S AFAR DEPRESSION GOT SUDDENLY WIDER IN LATE 2006 WHEN THIS 8 METRE WIDE EARTHQUAKE FISSURE RIPPED OPEN THE DESERT.**

plate motions. Its amalgamation appears to have been preceded by that of Pannotia about 650–550 million years ago, and the supercontinent Rhodinia existed around 1.8 billion years ago. Before that, at intervals of roughly 500 million years, the mythical-sounding lands of Nuna, Kenorland and Ur assembled and broke up in disorderly fashion, leaving only faint traces of their passing. Geologists believe that supercontinents, much like political superpowers, are always destined to break up. That's because their enormous girth traps mantle heat below, and as the underlying mantle overheats cracks appear in the crust and magma escapes, gradually wrenching the landmass open.

Amazingly, we can actually witness a continent being torn apart today. In September 2006, hundreds of earthquakes rocked a stretch of the Afar Depression to the south of Erta Ale, causing a minor eruption of one volcano (and reports of outpourings at Erta Ale itself), and splitting open a fissure that ran for 60 km (37 miles) and gaped up to 8 metres (26 feet) wide. Although dramatic, these were a mere sideshow to an even more remarkable event taking place below ground. A few kilometres beneath the desert, some 2.5 cubic kilometres (0.6 cubic miles) of magma – twice the volume of material thrown out of Mount St Helens in 1980 – was being injected as a vertical sheet into the Ethiopian crust at a speed one and a half times faster than the flow rate of Niagara Falls. In an instant, the Afar Depression became 2–4 metres (6.5–13 feet) wider. Another step had been taken in the process that will ultimately, tens of millions of years hence, wrench eastern Africa away from the rest of the continent, allowing the sea to flood into the Great Rift Valley and cover the deserts and grasslands with muddy slime.

Land Ahoy!

In July 1831, something was disturbing the sea off the south coast of Sicily. The Mediterranean waters churned and boiled, spouting great bursts of steam and ash high into the air, followed by thunderous eruptions of fire and cinders. The stink of sulphur hung in the air and, all around, dead fish floated on the surface. Far below, subterranean volcanic eruptions were pushing up a speck of sea bed until it broke through the waves to become the Mediterranean's newest island, and a source of general astonishment.

It didn't take long for this tiny scrap of rock, less than 2 km (1.2 miles) wide and 60-odd metres high, to cause an international political spat. The British saw its strategic value, close to major shipping lines, and promptly planted a flag on it, naming it Graham's Island after Sir James Graham, then First Lord of the Admiralty. No sooner had the British departed than it was claimed by King Ferdinand II of the Two Sicilies (Italy was not yet united) and renamed after him: Ferdinandea. The French and Spanish showed interest too, and the island became something of a tourist attraction. The diplomatic furore died only when everyone realized that, during the months in which they had been fighting over its sovereignty, the island had been sinking sheepishly back beneath the waves. By December it had disappeared.

In recent years, Ferdinandea has been up to its old tricks again. In 1987 it was sufficiently close to the surface to be mistaken for a Libyan submarine by US aircraft and unceremoniously depth-charged. By February 2000, fresh eruptions had brought the summit of the volcano to within 8 metres (26 feet) of the surface – a threat to shipping. Fresh volcanic activity a year later led geologists to suggest that the next eruption would bring the island back above the sea, reopening those old but unresolved diplomatic wounds. Meanwhile, the Italians have reportedly had a diver plant another flag on this subterranean territory to remind rivals who owns it. We must simply wait to see when it rises again.

BELOW **RECENT ERUPTIVE ACTIVITY ON THE UNDERWATER VOLCANO OF FERDINANDEA SUGGESTS THAT THIS RISING LUMP OF MEDITERRANEAN SEA BED MAY BE ABOUT TO EMERGE AS A NEW ISLAND. IF IT DOES, WHO WILL OWN IT?**

ABOVE **ICELAND'S THERMAL SPRINGS ARE FABULOUS FOR RELAXING IN, PROVIDING YOU DON'T MIND SHARING YOUR BATH OR WORRYING ABOUT WHAT'S BENEATH YOUR FEET.** OPPOSITE **THE REGULAR TANTRUMS OF STROKKUR GEYSER IN ICELAND ARE STIRRED BY THE SAME CONVECTIVE FORCE THAT DRIVES THE PLANET.**

And what is happening in East Africa is going on elsewhere. The world over, continents are moving, colliding and splitting, oceans are widening or closing. The evidence, written in the rocks below our feet, allows us to reconstruct how Earth's face has changed through time, but we can also run the tape forward and predict what will happen in the future. In the Mediterranean, the motley crew of lands trapped in the converging vice of Africa and Europe will get crumpled up and pushed higher, eventually equalling the Himalayas. Australia is on a collision course with Asia. Already it is ploughing through the islands of Southeast Asia as it forces its way north. Projecting a few tens of millions of years into the future, its left shoulder will get caught and it will swing anticlockwise to embed itself into Borneo and southern China, in the same way that India collided with Asia 50 million years ago. Meanwhile, the Americas will continue to drift away from Africa and Europe as the Atlantic Ocean steadily widens, but eventually a subduction zone will form on one side of the ocean, and the ocean floor will begin to plunge back down into the mantle. The widening will stop, the Atlantic basin will start to shrink, and the Americas will come smashing back into 'Euro-Africa'. And so, 250 million years from now, a new supercontinent will be born – just one more stage in a never-ending cycle of planetary makeover driven by Earth's heat engine.

Fire and Ice

Every five to ten minutes, a jet of boiling water is fired some 20 metres (70 feet) into the air. After a few seconds, the fountain subsides and the waters around it calm down, the pool regaining a sheen of placid innocence. And then it jets again. This is Strokkur, also known as 'the great gusher', one of the most powerful geysers in the world and a must-see tourist attraction on the outskirts of Iceland's capital Reykjavík. It lies only a few metres from Geysir, the original geyser, though the latter now has only infrequent spitting fits and is often dormant for years at a time. Strokkur is the crowd pleaser, and an endless succession of visitors stream past to gasp at its regular tantrums. But what few of them realize is that the same force that drives this geyser, or indeed the equally famous 'Old Faithful' in America's Yellowstone National Park, also drives the planet.

Icelanders are all too familiar with the capricious nature of Earth's volcanic actions; no place better reflects the volatility of our planet. Earthquakes, floods, glacial bursts, sea surges, storms and climate chaos all afflict this tiny speck of land, and for more than 1000 years generations of the island's tough community have had to confront regular outpourings of lava, some of them extraordinarily violent, a few of them catastrophically lethal. But for every yin there is a yang. Many homes are fuelled by power plants that feed off the hot waters circulating below ground, and the thermal springs are fabulous for relaxing in, providing you don't mind sharing your bath with the hundreds of tourists who flock here to enjoy them. They come because the mineral-rich waters are meant to have therapeutic qualities, but the rollicking bathers might not be so relaxed if they knew exactly how the water was being heated.

Iceland sits atop a 'hotspot' – an enormous plume of rising hot rock, up to 100 km (62 miles) in diameter, that acts rather like a colossal Bunsen burner deep inside Earth's mantle. About 200 km (125 miles) below the island, the upwelling mantle plume spreads out,

searching for ways to break through to the surface. Molten rock squeezes up through a swarm of fissures, akin to those in the Afar Depression, where the land has been prised open and depressurized by a tug-of-war between the tectonic plates of Europe and North America, and the heat bursts out as volcanoes, geysers and bubbling mudpools. The effects of the tectonic motion are easy to see in the north, around Krafla volcano. Every few months here, cracks in the ground widen and new ones appear. Sometimes the ground rises by one or two metres before abruptly dropping, signalling an impending volcanic eruption. Between

1975 and 1984, this part of northeast Iceland widened by 7 metres (23 feet), nudging New York and London further apart in the process. The tectonic schism between east and west has even shaped the nation's history. In the year 930 AD, a group of Icelandic chieftains gathered at Thingvellir in southwest Iceland to sort out their disputes. They met under a towering rock wall that forms a natural amphitheatre with fantastic acoustics, and for the next eight centuries this was the seat of Icelandic government. What those early pioneers of democracy didn't know was that they had set up their parliament in the centre of the volcanic rift between the North American and European plates – an auspicious site for the world's first parliament.

ABOVE **THE WORLD'S OLDEST DEMOCRATIC PARLIAMENT USED TO MEET IN THE OPEN AIR HERE AT THINGVELLIR, ICELAND, A NATURAL VOLCANIC CLEFT WHERE THE NORTH AMERICAN AND EUROPEAN PLATES MEET.**

Iceland's giant plume of magma doesn't just heat the island – it also created it. Molten lava has been spewing on to the Atlantic floor here for millions of years, so it should have been no surprise when, in 1963, the ocean began to boil and red-hot lava appeared in the

ABOVE **A SATELLITE VIEW OF ANAK KRAKATOA VOLCANO.** OPPOSITE **THE FIERY LANDSCAPE OF MOUNT KILAUEA ON HAWAII'S 'BIG ISLAND'.** OVERLEAF **THE LEAKY PLUMBING OF KILAUEA REGULARLY DELIVERS RIVERS OF RED-HOT LAVA TO THE OCEAN.**

swell around 100 km (62 miles) southeast of Reykjavík. The fresh lava hissed into the ocean for three and a half years, building layer upon layer into an island that was soon the second largest in an archipelago collectively known as Vestmannaeyjar. This chain of islands harbours Iceland's fishing fleet, so it was a cause of national concern when, a decade later, a second burst of volcanic unrest erupted. This time it was on the inhabited island of Heimaey, the largest of the Vestmannaeyjar chain. Heimaey is only a fraction of the size of Manhattan, yet a mile-long fissure was pouring out enough lava to bury New York's entire financial district and leave only the tops of a few skyscrapers sticking out. And the eruption was happening right on the doorstep of Heimaey's town and harbour, which was also Iceland's most important fishing centre, producing about a twelfth of the country's export income. To lose this would be a national disaster. The town was evacuated by fishing boats in the night, but several hundred locals returned to fight back the tide of lava with water pumps and bulldozers. Working 48-hour shifts, they fought for months to stem the magisterial advance of the lava front and halt its assault on the town. On 3 July 1973, the eruption ended and the town was saved, its precious harbour intact.

Powered by magma plumes, volcanoes have created islands around the world. While Iceland sits astride a plate boundary, other hotspots occur where plumes have broken through the middle of ocean plates. One of the best known is in the centre of the

ABOVE FOR SOCIETY, EARTHQUAKES ARE FORCES OF DESTRUCTION (AS SEEN HERE IN THE CITY OF KOBE, JAPAN, IN JANUARY 1995), BUT FOR THE PLANET THEY ARE FORCES OF CREATION AND RENEWAL.

Pacific Ocean, where a giant upwelling of hot rock has punched repeatedly through the moving plate above. Today, the plume sits beneath Hawaii, stoking Mount Kilauea – the world's most active volcano – but stretching to the northwest is a chain of extinct volcanoes that sat over the plume in times past. The largest of these still project above the sea and form the Hawaiian archipelago, but there are dozens more, increasingly old and more eroded, that have sunk from view and form a vast chain of submarine volcanoes: the Emperor Seamounts. But the island of Hawaii is where the action is now. Measured from their roots on the sea floor, the Hawaiian volcanoes Mauna Loa and Mauna Kea are the largest and tallest mountains on Earth, some 200 km (124 miles) wide at the base and more than 12 km (7.5 miles) tall. And with Mount Kilauea continually pouring out new lava, Hawaii's 'Big Island' is still growing. But there is a pretender in waiting. Just to the southeast sits a young submarine volcano, called Loihi, that is destined to grow into the next Hawaiian island.

The vast majority of Earth's volcanoes are found at plate boundaries, which criss-cross the planet like the seams of a baseball. Many thousands seep silently on the ocean floor, lining the seams where plates pull apart, and only rarely coming to our attention. But most of the biggest and most violent volcanoes are on land and are found where plates are pushing together rather than pulling apart. As one plate is forced beneath the other, the tremendous heat and pressure deep underground causes rock to melt, and some of the molten rock finds its way back to surface, building into active volcanoes such as those around the Pacific's so-called ring of fire.

Moonquakes and Marsquakes

Earthquakes don't just happen on Earth – though, understandably, geologists prefer to give the extraterrestrial versions different names. The Apollo astronauts who visited the Moon in the early 1970s installed earthquake-detection instruments – seismometers – and four of these successfully recorded 'moonquakes' until the late 1970s. Unlike Earth, our Moon does not have tectonic plates. However, it does suffer from shallow tremors caused by the heating and cooling of its surface, and from deeper shocks, 700–1200 km (400–750 miles) underground, caused by Earth's tidal forces. Just as the gravitational pull of the Moon causes our ocean tides, so the much more powerful gravity of Earth tugs away at the Moon's crust.

Elsewhere in the solar system, the situation is not so clear. The *Viking* spacecraft took a seismometer to Mars, and it operated for more than five months. It recorded several possible 'marsquakes', but they could have been caused by wind buffeting *Viking*'s landing apparatus. Mars may have had moving tectonic plates in the past, but there are no signs of them now. Instead, our planetary neighbour seems to have what's called a 'stagnant lid' – a strong shell that keeps the interior too hot for plate tectonics to work but occasionally lets magma break through, blistering the Martian surface with enormous volcanoes. Some of the surface features observed on radar pictures of Venus resemble earthquake faults on Earth, but the lack of water in the Venusian crust ensures that plate tectonics is impossible there. Crater-count studies suggest that the surface of Venus had a complete makeover 300–600 million years ago, but there is no evidence of any movement since then. Further afield, Jupiter's moons Io and Europa look as though they might suffer from seismic shocks, possibly caused by Jupiter, whose enormous mass must create tidal forces far greater than those that Earth inflicts on our Moon. But for now we have no seismometers to make sure.

BELOW **MARS MAY HAVE HAD A MOBILE CRUSTAL SKIN IN THE DISTANT PAST, BUT A LACK OF 'MARSQUAKES' IMPLIES THE FACE OF THE PLANET IS STOLID.**

Volcanoes in Japan, Indonesia, the Philippines, the Aleutian islands and along stretches of the west coasts of North, Central and South America include some of the most active and best known in the world.

The Earthquake Cycle

In 1835, the English naturalist Charles Darwin was midway through the five-year voyage on the *Beagle* that took him to the Galápagos Islands and helped inspire his theory of evolution. But what preoccupied Darwin for much of his journey up the rugged Chilean coast of South America wasn't the origin of species but the origin of the spectacular geology around him. It has been said that it is the natural and legitimate ambition of a properly constituted geologist to see a glacier, witness an eruption and feel an earthquake. Well, by the time Darwin had reached the volcanic hotspot of the Galápagos archipelago, he was well and truly a geologist, for he had just witnessed the majesty of all three.

Having left the ice fields of Patagonia, where he strode among glaciers whose tongues came right down to sea level, Darwin arrived in central Chile to find the mountainous coast bedevilled by seismic and volcanic unrest. On 19 January, with his ship anchored in a calm bay on the island of Chiloé, he witnessed a magnificent midnight fireworks display erupting from the mainland volcano of Orsono. It was a busy night across the region: a large earthquake shuddered, and two other volcanoes erupted – Aconcagua, 770 km (480 miles) to the north, and Coseguina, 4300 km (2700 miles) further north in Nicaragua. In terms of the vast distances involved, it was, Darwin wrote, equivalent to the simultaneous eruption of the Italian volcanoes Vesuvius and Etna and Iceland's Hekla. A month later, on 20 February, came the biggest earthquake in generations. Up and down the coast that he had travelled, harbour towns were instantly levelled; in some, not a house was left standing. And there was more misery to come. A great wave – a tsunami – appeared off the coast, visible from miles away as a terrifying line of white breakers. It swept ashore with irresistible force, rushing 7 metres (23 feet) above the highest tide mark and overwhelming those emerging from the carnage of the quake. And once again, across the region, volcanoes burst forth.

The intensity of the geological mayhem was astounding, but equally remarkable was its vast reach. To convey the scale of the phenomenon, Darwin imagined the same thing happening in Europe:

> *Then would the land from the North Sea to the Mediterranean have been violently shaken, and at the same instant of time a large tract of the eastern coast of England would have been permanently elevated, together with some outlying islands – a train of volcanos on the coast of Holland would have burst forth in action, and an eruption taken place at the bottom of the sea, near the northern extremity of Ireland – and lastly, the ancient vents of Auvergne, Cantal, and Mont d'Or would each have sent up to the sky a dark column of smoke, and have long remained in fierce action. Two years and three-quarters afterwards, France, from its centre to the English Channel, would have been again desolated by an earthquake and an island permanently upraised in the Mediterranean.*

Darwin experienced the January earthquake's amazing effects first-hand. The land around the bay in which the *Beagle* was anchored was raised by about a metre (3 feet), while more distant islands jumped up 2–3 metres (6–10 feet), leaving beds of putrid mussels high and dry. On his frequent walks along the coast, Darwin found equivalent mussel beds at elevations of several hundred metres. To this extraordinary man, the implication was clear: the great height of the Andes

OPPOSITE **THE FURROWED TRACE OF THE SAN ANDREAS FAULT, A DEEP TECTONIC SCHISM THAT SEPARATES WEIRD AND WONDERFUL COASTAL CALIFORNIA FROM THE REST OF NORTH AMERICA.**

mountain range was the sum of successive small risings, its elevation reached by thousands of sudden jumps of the coast, similar to the one he had just lived through. Charles Darwin, the naturalist, had just discovered the 'earthquake cycle'.

According to the earthquake cycle theory, violent tremors happen when stored energy is released suddenly at faults – the weak hairline fractures in Earth's crust. The theory emerged out of the smouldering ruins of San Francisco in 1906, after a great jolt set the city alight and shunted California's Pacific coast several metres northwards relative to the rest of America. The trail of wreckage from the 1906 quake – including scarps, fissures in the ground, split trees, and wrecked homes and roads – revealed a previously hidden crack in the crust that stretched for 100 km (62 miles). At the time, few geologists thought faults caused earthquakes, most preferring to blame subterranean explosions of pent-up gas, and some even clinging to the notion of divine wrath. But after the 1906 'fire', as local officials referred to it (not wishing to dampen the economic spirits of the gold rush city with thoughts of seismic fallibility), faults were exposed as the true culprits. Today it seems obvious, with the San Andreas Fault being one of California's most distinctive natural landmarks. Its path is clearly visible from the air – on a clear day a pilot with no radio or GPS can fly the length of the state navigating only by its distinctive furrow, and even from space it appears to knit California together like a surgical suture.

In the modern parlance of plate tectonics, faults are where the slow, continuous drift of the plates – powered by the equally inexorable creep of hot rock underneath – becomes snagged. The friction of huge sheets of rock grinding against each other locks the movement along an intervening fault. Even though strains continue to pile up on it, the fault sits still for a time. The pressure builds, the strength of the rocks is overcome and finally the fault jumps. The land opens like a zipper, and a fault break rips along the ground like a runaway locomotive, unleashing seismic energy in the form of the shaking waves that radiate out from the point of failure. Then all goes quiet again and the fault returns to obscurity until prompted to slip again.

Earthquake zones encircle the globe, marking out the tectonic boundaries where plates split and pull apart, grind alongside each other, or converge. The great earthquake that Darwin witnessed in Chile was just one of many that shake the western coast of South America, where the Pacific sea floor is being pushed under the buoyant continental plate, one great sheet of rock thrusting under another in a series of enormous seismic jolts. From South America, swarms of earthquakes follow the chain of volcanoes around the Pacific Ocean's ring of fire to Southeast Asia, then through Micronesia to New Zealand. In Southeast Asia we meet another of the world's great seismic belts, which extends west through China, the Himalayas, the Middle East, southern Europe and out through the Strait of Gibraltar. These have all been sites of great earthquakes, such as the 1755 catastrophe that levelled bustling Lisbon; the deadly 1999 earthquake that fired a warning shot at Istanbul; and the calamitous 2004 Boxing Day quake off Sumatra that radiated tsunami devastation around the Indian Ocean. It is usually destruction that comes to mind when we think of earthquakes, but this is our skewed human viewpoint – as Charles Darwin realized a century and a half ago, earthquakes are also forces of creation and renewal.

How Erosion Builds Mountains

'Mountains', mused the Victorian artist and writer John Ruskin, 'are to the rest of the body of the Earth, what violent muscular action is to the body of man.' In the mid-nineteenth century, when Ruskin was writing, mountains were thought to derive from a convulsive action of Earth's outer skin, the land wrinkling as it contracted like the skin of an old apple. By the early twentieth century, mountain ranges instead owed their prominence to long troughs of thick sediment ('geosynclines') being squeezed shut and crumpling upwards. Today, modern geology

Sculpted by Rain

Limestone is one of the most common types of rock on the surface of the planet. Most of it forms in shallow seas as a result of the unceasing death and burial of organisms rich in carbon carbonate. Over millions of years, their fossilized corpses pile up to form layers of limestone that can be kilometres deep – a vast, submarine reservoir of carbon that was once warming the atmosphere but is now locked away, trapped in solid rock. In places, Earth's shifting plates have lifted these limestone sea floors above the waves to become land, exposing the rock once again to the weathering power of the atmosphere. Rain corrodes limestone far more easily than other types of rock, often sculpting it into strange and spectacular landscapes.

One of the most dramatic limestone landscapes is in northern Madagascar, where serrated towers and pinnacles of ancient calcium carbonate peek above the jungle treetops. These strange formations are known as *tsingys*, which in the local language means 'tiptoe', and as soon as you walk among them you can see why: the fluted edges of the pinnacles are like razors, sharpened by the aggressive tropical rain. At whatever scale you look – whether you're scrutinizing the shell-rich surface with an eyeglass or peering down from a helicopter – you see the pale grey bedrock riven with pits and runnels, cracks and fissures, all of which attest to the slow and remorseless etching away of this ancient sea bed.

BELOW **IAIN STEWART TIPTOEING THROUGH THE JAGGED LIMESTONE PINNACLES OF MADAGASCAR'S *TSINGYS*.**

views mountains as the planet's most spectacular expression of the creeping plates that remorselessly squeeze, bend, break, shift and uplift rocks. Earth's three main belts of mountains all define major plate boundaries: the Alpine and Himalayan peaks form the crumpled southern margin of Europe and Asia; the circum-Pacific ranges follow the subducting edges of the ring of fire; and the rugged shoulders of East Africa span the continent's unzipping rift valley. Mind you, some of the biggest mountains on the planet aren't on land at all.

Mountains Under the Sea

If we could pull the plug on the oceans and drain away the water, we would see a truly awesome sight. Held aloft by its underlying mantle plume, Iceland would stand out as the highest peak of the longest mountain range in the world. This is the Mid-Atlantic Ridge, the giant backbone that runs for more than 15,000 km (9300 miles) beneath the Atlantic Ocean, rising as much as 4 km (2.5 miles) above the surrounding abyssal plains of the ocean floor. Similar mid-ocean ridges run from the Atlantic into the Indian Ocean and across the Pacific, all marking places where plates are moving apart and new crust is building up. The result is 60,000 km (37,000 miles) of more or less continuous mountain chain. No other planet in the solar system has such a feature; linear mountain chains like this are Earth's and Earth's alone.

Even ancient mountain ranges, now nestling in the deformed hearts of continents, thousands of kilometres from the nearest mid-ocean ridge or subduction zone, are the remnants of once clashing plates. Retired tectonic belts like the Appalachian Mountains of eastern North America, the Caledonian mountains of Scotland, Greenland and Svalbard, and the Ural Mountains of Russia simply mark those places where the action used to be. But these uplands are hundreds of millions of years old, so why are they still upstanding? Surely they should have been long since dispensed with by erosion? It turns out that although you need tremendous pressures to build mountains, you need equally strong forces to get rid of them.

Rocks and mountains might seem solid and permanent, but stroll through any old cemetery and you'll see stone monuments with crumbling edges and illegible epitaphs worn down by barely a century or two of rain, sun and wind. The planet doesn't just have centuries at its disposal, it has tens and hundreds of millions of years, and over these timespans whole mountain ranges ought to disappear. What's more, mountains start to disintegrate as soon as they rise – they literally come apart at the seams as pressure is relieved by the removal of overlaying rock layers. Weaknesses in the rock begin to open up, admitting air and water, and thus begins the process commonly known as weathering. Weathering can be mechanical, where rock is physically broken into fragments, or it may happen chemically, with air and rain reacting with the minerals in rock to form soft residues that are easily stripped away. It happens imperceptibly slowly and is often hidden from view, but it is nature's first step in wearing down mountains. The next step is more visible: erosion strips away the weathered debris, sweeping it out of the mountains and, ultimately, out to sea. For centuries geologists thought this too was a slow and gradual process, but in mountain ranges it can happen with brutal swiftness.

Shortly after midnight on 14 December 1991, climbers sleeping in a hut beneath the east face of Mount Cook – New Zealand's highest peak – were woken by a low rumble that quickly became a loud roar. Looking up, they saw orange sparks flying in the dark as rocks began crashing down. Within seconds, a maelstrom of ice and rock swept past them at several hundred kilometres per hour, travelling as a watery,

OPPOSITE **THE HIMALAYAS, THE 2700 KM (1700 MILE) LONG MASSIF THAT WAS FORMED WHEN INDIA COLLIDED WITH EURASIA. IN THIS FALSE-COLOUR SATELLITE IMAGE THE RED AREAS INDICATE VEGETATION.**

ABOVE **THE SPIKY PEAKS OF NEW ZEALAND'S SOUTHERN ALPS ARE RISING THANKS TO THE EROSIONAL ACTION OF GIANT ROCKSLIDES, CREEPING GLACIERS AND RAGING RIVERS.**

turbulent mass. When dawn broke, the scale of the night's events became clear. Some 14 million cubic metres (18 million cubic yards) of rock from the summit lay strewn as rubble and dust across the Tasman Glacier and the river valley beyond. In an instant, the peak of the country's greatest natural icon had been lowered by 10 metres (33 feet). Relief that had taken thousands of years to build had been destroyed in a matter of minutes.

Geologists have a word for the serrated skyline of New Zealand's mountain backbone, the Southern Alps; they simply call it 'spiky'. Rising by up to a centimetre (0.4 inches) a year, these Alps form a barrier 2–4 km (1–2.5 miles) tall in the path of the prevailing, moisture-laden winds that blow off the Tasman Sea. As much as 15 metres (590 inches) of rain is dumped annually on the steep, western slopes, and the water seeps into the ground and eats away at the rock. The rock is weak and rotten. An ancient, sandy sea bed that was remoulded by heat and pressure to form fissile vertical strata, it has been shattered by earthquakes, prised apart by freezing and thawing, and corroded by rain and air. It collapses readily, making the jagged, precipitous peaks notoriously dangerous to climb. Mountaineers joke that if you are scaling these crumbling crags you should take a pebble in your pocket – that way, you always have something solid to hold on to. All along the ragged spine of this rugged country, the splintered crests are falling away. Landslides on their own are capable of lowering the peaks as fast as subterranean forces thrust them skyward, and the bulk of their erosional labour is carried out by infrequent but giant rock avalanches like the one that still scars Mount Cook. But other denuding forces are at work here too. Mountain glaciers aggressively gnaw the rock, grinding down the summits that project above the snowline. Far below, whitewater rivers thrash through deep, V-shaped gorges carrying rafts of squealing backpackers as readily as mountain detritus. Everywhere there is motion, everywhere there is change.

Rockslides, glaciers, and rivers pick away remorselessly at the bones of the Southern Alps. Their flesh pared by erosion, these are made lighter by the removal of mass from their vast bulk, and so the dissected skeleton rises. Mountains are like icebergs: their rigid crustal bulk floats on a warm, pliable mantle below, with most of their mass hidden below the surface and only their tips projecting above. Erode the top off an iceberg and it just bobs higher in the water, adjusting its position so that it maintains nine-tenths of its mass below the waterline. Geologists call this tendency 'isostasy', and it is by isostatic uplift that mountain ranges maintain their relief. Mountains are not somehow created whole and subsequently worn away – they wear down as they come up, rising and shedding debris steadily through time.

The Himalayas

Nowhere displays this better than the most majestic mountains on the planet: the Himalayas. The centrepiece of the range is Mount Everest, which, at 8848 metres (29,028 feet) high, is 15 times taller than anything humans have ever built. It sits among the 100 highest mountains on Earth, which form a vast arc stretching for 2700 km (1700 miles) and separating India and Nepal from Tibet with one of Earth's most dramatic barriers. Every year, moisture-laden monsoon winds from the Indian Ocean sweep north across the Indian subcontinent and the Bay of Bengal to slam into the ramparts of the Himalayas. As much rain is dumped on the slopes of northern India and Nepal in the monsoon as falls in the Amazon basin in a whole year. Thunderhead rain clouds struggle vainly to climb the Himalayan peaks, but little rain or snow makes it over the 'roof of the world' and on to the Tibetan plateau beyond. The resulting rain shadow creates one of the most dramatic climate contrasts on the globe: from lush tropical forest on the southern flanks, through dizzy alpine peaks and into high-altitude desert – all within a matter of miles.

South of the rain shadow, the torrential monsoonal deluges feed some of the highest erosion rates on the planet. The violence of the rain loosens snow and ice from the peaks, and thundering avalanches deliver glacial debris directly into bloated, milky-white headwaters, which twist and plunge through cascades and cataracts that bite deeply into the surrounding mountains, ultimately destined for a more leisurely course down the Ganges or the Brahmaputra to the ocean. In the rugged hinterland, the enormous amounts of rock liberated by the relentless raging torrents over several million years has allowed the steady isostatic rise of the mountain tops, while at the same time luring deeper-seated rocks closer to the surface. Slowly but surely, the roots of the mountain belt, formed tens of kilometres below the ground, emerge, grudgingly, along the axis of greatest erosion.

Other young mountain belts, like the Andes, the Cascade Range of western North America, and the coastal ranges of Taiwan, follow the Himalayas and the Southern Alps in rising and eroding in fairly even gait. Even ancient uplands are not immune. After hundreds of millions of years, the rolling hills of the eastern USA's Appalachian highlands are still rising by a few millimetres a year, courtesy of the rivers that are cutting down through them, and it is a similar story for other ancestral ranges. This isostatic seesaw ensures that, although mountains are far from permanent, they can stay standing for astonishingly long, their heads held high by the very processes that eat away at them.

But perhaps even more remarkable is the discovery that the drumbeat of raindrops on mountains turns out to be a crucial part of the mechanism that controls our planet's climate.

Supervolcanoes

They pack as much punch as a mile-wide asteroid but strike the Earth ten times more often. It is only in the last decade or so that the spectre of supervolcanoes has come to light, partly because the stupendous eruptions they can generate have never happened in recorded history, and partly because these behemoths are mostly hidden below ground.

Supervolcanoes are scattered across the globe, but thankfully there are few of them. The largest lurk beneath Yellowstone National Park in Wyoming, Long Valley in California, Lake Toba in Sumatra and the Taupo district of New Zealand.

They are thought to form where a vast, swollen pool of hot mantle rock has punched into the overlaying crust with such force that fractures develop, allowing magma to surge upwards through a network of vents. The vents link up in a ring, causing the central block of crust to lose its structural integrity and cave in. The collapse depressurizes the magma chamber below, and the result is like popping the cork out of a champagne bottle: suddenly, virtually the entire reservoir of magma explodes outwards in a cataclysmic blast. Scalding volcanic fumes and ash are hurled all the way to the stratosphere at supersonic speed, injecting billions of tonnes of debris and gas into the sky. A veil of smog cloaks the planet, leading to a 'volcanic winter' with the potential to bring about mass extermination of life. In the long term, volcanic gases such as sulphur dioxide and carbon dioxide cause global warming, and recent studies indicate that the noxious fumes may also eat holes in the ozone layer, allowing deadly levels of ultraviolet radiation to reach Earth's surface.

Only volcanoes capable of storing massive reservoirs of magma can aspire to supervolcano status. Members of the elite club must have the ability to release at least 300 cubic km (72 cubic miles) of magma – enough material to bury all of Greater London to a depth of 200 metres (650 feet). And remember, that is a baby supereruption. Some 35,000 years ago in the Bay of Naples, Italy, an event of this size blanketed the surrounding Campania region with metres to tens of metres of ash. Some scientists claim a consequent deterioration in the global climate was the final straw for the Neanderthals, who were already in terminal decline due to the encroachment of modern humans.

Our own species may nearly have succumbed to a similar fate. About 74,000 years ago, a supervolcano erupted on the site of Lake Toba in Sumatra. It was ten times larger than the Neapolitan blast, hurling out 3000 cubic km (700 cubic miles) of magma and creating a zone of death and destruction thousands of kilometres wide. Evidence from ice cores suggests the resulting volcanic winter lasted many years and caused a global ecological crisis. Some argue that modern humans, who had left Africa only 30,000 years before, came close to being wiped out: genetic data suggest there was a population 'bottleneck', with our entire species winnowed down to perhaps only 10,000 individuals. To some anthropologists, the cooperative social and communication strategies that modern humans later employed to survive the ice age, and perhaps to outcompete the Neanderthals, may have been survival mechanisms that developed during this global volcanic winter. According to that still-controversial view, we are descendants of the few small groups of tropical Africans who united in the face of adversity.

BELOW **TOBA IN INDONESIA IS THE BIGGEST LAKE IN SOUTHEAST ASIA. BUT DON'T BE FOOLED BY ITS SERENITY – BENEATH LURKS ONE OF EARTH'S BIGGEST SUPERVOLCANOES.** OPPOSITE **A FROTHING VOLCANIC MUDPOOL AT ROTORUA IN NEW ZEALAND.**

Global Thermostat

It is a drop of rain that sets in motion Earth's amazing temperature-control mechanism. When rain falls, it absorbs carbon dioxide from the air – there's a tiny bit of the gas dissolved in every drop of water that falls. The two combine chemically, forming a weak acid: carbonic acid – the substance that gives carbonated drinks their tang as well as their bubbles. Carbon dioxide is reluctant to dissolve without pressure, so rain contains far less of it than a fizzy drink and is much less acidic. Even so, it has enough bite to attack the silicate minerals in solid rock over long periods of time, turning them into soft and crumbling clay. This chemical reaction is the basis of weathering, and it has two very important consequences: it destroys rock, and it takes carbon dioxide out of the atmosphere.

The carbon dioxide ends up chemically trapped in water as bicarbonate, which washes away in rain, streams and rivers, and is eventually flushed out to sea. And that's where life gets in on the act. Underwater, millions of tiny sea creatures like plankton, as well as bigger marine organisms like corals, combine the bicarbonate with calcium from sea water to make calcium carbonate, a hard mineral that's perfect for shells and skeletons. When they die, their hard body parts fall to the sea floor as mud, taking the carbon with them. The mud hardens over time, compacted by the weight of fresh layers falling on top and cemented by minerals crystallizing within it. And so it slowly turns into carbonate rock: limestone. Locked securely within it is the carbon that once fell to Earth in rain.

Where ocean floors buckle under the pressure of converging tectonic forces, limestone on the sea floor can be lifted up to form land or even mountains (*see* 'Sculpted by Rain', page 91). However, most of it is shunted along the tectonic conveyor belt towards a subduction zone under the coast of one of the continents. Here, the ocean crust disappears down the hatch, descending into the hot mantle. Tens of kilometres below the surface, a remarkable transformation occurs. The mass of fossilized sea creatures becomes squeezed and heated to such a degree that the limestone melts and its carbon is liberated, turning back into carbon dioxide in the resulting magma. Buoyed by its gassy constitution, this molten rock creeps up through fractures and fissures until it eventually reaches the surface. And there, it blasts out of volcanoes.

Volcanoes, those seething cauldrons of vapour, constantly pump out greenhouse gases into the atmosphere, forming a veil around Earth that traps heat from the Sun. Today we think of greenhouse gases as a bad thing, causing havoc through climate change, but these insulating fumes are absolutely vital to our planet – without them, Earth would have an average surface temperature of −20°C (−4°F). However, if volcanoes were left unchecked, the continual outpouring of greenhouse gases would cause Earth to overheat. It is humble old weathering that saves us from this fate by removing carbon dioxide from the atmosphere and sending it back to the muddy sea bed, to be cycled yet again through volcanoes.

What's more, the system works like a thermostat, keeping Earth's temperature within a comfortable range. When the climate gets warmer, the rate of weathering increases, drawing more carbon dioxide into the ocean and reducing the greenhouse effect; and so the climate cools. This cooling then reduces the rate of weathering, allowing carbon dioxide from volcanoes to build up again; and so the climate warms. It is a remarkable natural balancing act, and at its heart is Earth's inner heat engine. It is the heat from within that drives the ever-moving crust, pushing limestone-draped sea floors towards volcanoes to liberate the trapped carbon and return it to the air. And those same motions create the mountain ranges that provide an endless supply of fresh rock to be weathered. It is the ultimate recycling scheme – a remarkable, global thermostat that elegantly links the planet's atmosphere, ocean and hot interior to keep the temperature at the surface perfect for life.

ABOVE **GIANT'S CAUSEWAY IN ANTRIM, NORTHERN IRELAND – A LEGACY FROM THE PLANET'S MOST RECENT GARGANTUAN OUTBURST OF LAVA FLOWS 55 MILLION YEARS AGO.**

Volcanoes and Mass Extinctions

The notion that volcanoes somehow nurture and protect life might seem strange, given how dramatically violent and destructive we know they can be. In fact, it seems that on several occasions in our planet's history, volcanoes have tested both the global thermostat and the robustness of life to the limit (*see* 'Supervolcanoes', page 96).

Every so often, Earth can erupt truly colossal quantities of lava in a short time, producing what geologists call a large igneous province, or LIP for short. Formed from multiple gigantic floods of lava, LIPs consist of layers of the volcanic rock basalt stacked together, and they can cover an enormous area. The Deccan plateau in India, for instance, is made up of flood basalt up to 2 km (1.2 miles) thick covering an area the size of Spain. And these vast lava flows can form in only a few hundred thousand years – a geological blink of an eye.

Quite why such massive outpourings happen isn't clear, but they may involve the gargantuan throat-clearing of mantle plumes in the deep. In 1994, the French geologist Vincent Courtillot suggested in his book *Evolutionary Catastrophes* that these exceptional events were the trigger for mass extinctions – occasions when a significant proportion of Earth's species disappears suddenly from the fossil record. In the last 500 million years there may have been up to 15 such episodes, five of which eliminated more than half the species known to inhabit our planet. In the mid 1990s, theories about asteroid impacts were all the rage, but a decade on, accurate dating of LIP events suggests their timing corresponds much more closely with that of mass extinctions. Thus, while the single most devastating cull of life was taking place 250 million years ago, the planet was busy producing the largest igneous province that currently exists on land: the Siberian Traps in central Russia. ('Traps', from the Swedish word for stairs, refers to the steplike slopes that form when the basalt layers erode.) And while the dinosaurs were being exterminated 65 million years ago, enormous lava floods were forming the Deccan Traps in India; the asteroid strike on Mexico's Yucatán coast seems just to have exacerbated an already dire situation. The most recent flood basalt incident took place 55 million years ago, when a vast hotspot ripped open the sea floor in the northern Atlantic, suddenly warming the oceans by around 5°C (9°F) and triggering a major loss of marine life. Today, the western Highlands of Scotland are the unlikely setting for many of the volcanic features left behind by this catastrophe.

So, rare and extraordinary volcanic eruptions coincide with rare and extraordinary biological changes. No other phenomena – especially not the much-touted asteroid and comet impacts – have such a close correlation with extreme instances of ecological wipe-out. But how could copious eruptions of lava in one corner of the planet lead to drastic culls of life far away, both on land and at sea? Most geologists believe the answer lies in the gargantuan volume of carbon dioxide that such events release. As well as warming the atmosphere, the gas may somehow have asphyxiated the oceans by triggering a massive depletion of oxygen in the water – a so-called 'anoxic event'. This theory remains controversial, not least because the mechanism involved is unclear; but what is clear is that poisoning the atmosphere or the seas can have catastrophic effects – as we shall discover in the next two chapters.

RIGHT **STAYING AHEAD OF THE CLOUD – RACING TO ESCAPE AN ENORMOUS DELUGE OF ASH FROM THE 1991 ERUPTION OF MOUNT PINATUBO IN THE PHILIPPINES.**

ABOVE **THE 5140 TRILLION TONNES OF AIR IN EARTH'S ATMOSPHERE ARE REVEALED AS A TINY VOLUME WHEN DEPICTED AS A SINGLE SPHERE SUSPENDED IN ORBIT NEXT TO THE PLANET.**

After he reached maximum speed in the stratosphere, the gradually thickening air began to slow him down. Some 21 km (13 miles) below the gondola, he finally reached the troposphere and re-entered the world of weather. Now, with the deafening but reassuring roar of wind and cloud rushing past him, the sensation of speed was very great, and he fell about 5 km (3 miles) further before his main parachute opened at an altitude of 5.5 km (3.4 miles). From there on down, his descent was more stately and controlled, and he perhaps had a few moments to reflect on the records he had set, all of them still unsurpassed today. Among the things he recalls that struck him during the jump were the intense hostility of the outer atmosphere and the very thinness of the layer of air on which all our lives depend.

One thought that perhaps did not enter Joe Kittinger's mind as he neared the ground is how Earth's atmosphere and life are intertwined, for the air we breathe not only sustains life but was created by it. Since its very beginnings, life has changed planet Earth – sometimes subtly, sometimes dramatically – and the evolution of our atmosphere is perhaps the best example of the power with which living things shape our world. Such notions would not have occurred to Joe Kittinger because his daring leap took place well before talk of global warming or the ozone hole, and well before the famous Gaia theory proposed by the British scientist James Lovelock. It is only in the last two or three decades that it has become crystal clear to science how finely balanced Earth's life-support mechanisms are, and how the atmosphere in particular plays a central role in the complex web of connections that keeps our planet habitable. The balancing act involves intricate links between air, land, sea and life, with feedback loops and cycles that ebb and flow. It is a system that has been changing and evolving for 4 billion years, so to truly understand it we must return to the beginning.

First Breath

The story of our atmosphere is, in a sense, the story of a seesaw battle between two chemical compounds that are both vital to life: carbon dioxide and oxygen. The rise and fall of the one against the other in the atmosphere mark key stages in the story of the planet, highlighting critical steps in life's evolution.

Traces of the beginning of that epic duel can be found on the westernmost point of Australia, about a day's drive north of Perth. The trip will take you through some of the continent's most beautiful wild-flower countryside and will probably bring you face to face with a number of sheep that have wandered off unfenced grazing land, so it's wise to set off in time to finish your journey before dark. The coast here was once an active pearl-fishing area, and as you pass through the town of Denham you travel over a street paved with oyster shells. But your ultimate destination is Shark Bay: a magnificent stretch of shallow coastal waters populated by herds of dugongs – herbivorous marine mammals that feed on the world's largest sea-grass meadow. Some 10,000 foraging dugongs can be found swimming in Shark Bay, along with turtles, sea snakes, manta rays, bottlenose dolphins and humpback whales. But what is particularly special is the part of the bay that dugongs don't venture into. Hamelin Pool is a wide, shallow area with calm, crystal-clear water. Poking just above the surface, or shimmering just below it, are strange, knee-high lumps of what look and feel like stone, shaped a bit like the squat bollards

BELOW **AN INFRARED IMAGE OF THE ATMOSPHERE, REVEALING THE COMPLEX CURRENTS THAT TRANSPORT HEAT ACROSS THE GLOBE.**

CHAPTER THREE

Atmosphere

IN 1955, AT THE HEIGHT OF THE COLD WAR, THE US AIR FORCE BEGAN PROJECT Manhigh, a research programme designed to find out if humans would be physically capable of going into space. The project involved a series of very high-altitude balloon flights, and the first mission was undertaken by a young pilot called Joe Kittinger, whose seven-hour trip took him to an altitude of 29,500 metres (97,000 feet) before he was brought safely back to Earth. It was an extraordinary achievement, but for Kittinger it was merely a warm-up for the mission that would ensure his place in history.

At 2.00 a.m. on 16 August 1960, a ground crew began the long process of inflating a vast helium balloon on a disused airstrip in the desert of New Mexico. At 4.00 a.m., Joe Kittinger donned warm clothing and a pressurized suit and began adapting his breathing to pure oxygen, clearing nitrogen from his blood to reduce the risk of getting lethal decompression sickness – 'the bends' – if the pressure on his body were to drop too rapidly. He was kept air-conditioned throughout the warm desert morning to stop him sweating, so that his clothing would not freeze solid during the flight. At last, at 5:29 a.m., he climbed aboard an unpressurized gondola hanging beneath the balloon and rose, over a period of 100 minutes, to an altitude of 31,300 metres (102,800 feet) – a world record that remains unbroken outside of space flight.

OPPOSITE **THE THIN, HAZY VEIL OF THE ATMOSPHERE: WITHOUT IT, EARTH WOULD BE A DEAD PLANET.**

As he passed the 13 km (8 mile) mark, Kittinger discovered that the pressurized glove on his right hand was not working, but he pressed on, fearing the experiment would be aborted if he told ground control, and watched his hand swell to twice its normal size until he lost the use of it for the rest of his mission. High above Earth, at a temperature of –70°C (–94°F), Kittinger was in a world devoid of oxygen, with 99 per cent of the planet's atmosphere at his feet. Ahead he could see the curvature of Earth; above him there was nothing but blackness. He remained at this altitude for 11 minutes, and then, working carefully with his left hand, he disconnected himself from the power feeds of the gondola, checked his air supply, took a last look around, stepped to the doorway, pressed the start button on the cameras strapped to his body, said a small prayer – and jumped.

ABOVE **JOE KITTINGER LEAPING TO EARTH FROM THE EDGE OF SPACE.**
OVERLEAF **THE ATMOSPHERE MAY SEEM INVISIBLE AND EMPTY, BUT IN FREE FALL IT CAN BE SURFED LIKE ANY OCEAN WAVE.**

He recalls rolling on to his back and seeing the balloon shoot up into the blackness – though of course it was Kittinger who was shooting through the sky, accelerating to Earth at the start of what remains the highest and longest skydive ever undertaken. After 13 seconds, a small stabilizing parachute opened to prevent him going into a spin, but this was designed not to impede his fall, and by the time he had fallen 4 km (2.5 miles), he had attained a speed of 988 km/h (614 mph), or Mach 0.9 (90 per cent of the speed of sound). But for Joe Kittinger there was no sensation of speed, or sound, for he was falling though the stratosphere, that faint veil above Earth that forms one of the outer shells of our atmosphere.

We think of the clouds and sky as being far above us, an enormous depth of air that we and most other living things depend on for survival. But in reality, the atmosphere is surprisingly thin, measuring about 1 per cent of Earth's diameter – equivalent to the skin of an apple. It is made up of various distinct layers. The bottom layer, called the troposphere, is where all our weather happens – clouds, wind and rain are all confined to the troposphere. Yet this part of the atmosphere is a mere 10 km (6 miles) thick on average, and airliners can fly above it, giving stunning views across the cloud tops.

Next comes the stratosphere, where Kittinger started his jump. This layer extends to about 50 km (31 miles) high and contains the precious ozone gas that we now know defends life against ultraviolet radiation from the Sun. Temperatures fall as you rise through the troposphere, but in the stratosphere they rise again, stoked by the Sun's intense radiation. Above the stratosphere is the mesosphere, where shooting stars form, and above that is the thermosphere. The thermosphere includes the 'Kármán line' – the arbitrary dividing line between the atmosphere and space, defined, for purposes of convention, as being at 100 km (62 miles) altitude. In reality, however, there is no boundary and the atmosphere fades gradually over hundreds of kilometres. The few gas molecules in the thermosphere are easily split by the Sun's energy, forming charged particles called ions. Radio waves can be bounced off these, earning this part of the atmosphere the additional name 'ionosphere'. This is also the region in which the spectacular northern and southern lights reach closest to the surface. Finally, from about 700 km (430 miles) up, comes the highly rarefied exosphere, made up only of hydrogen and helium – gases so light they can escape into space.

Kittinger was in free fall for four and a half minutes.

The High Life

Depending on how you define its outer edge, Earth's atmosphere is anything from 100 km (62 miles) to thousands of kilometres thick. However, because of gravity, most of the air is at the bottom, and the zone of rich, breathable air is surprisingly small. About three-quarters of the total mass of air is in the bottom 11 km (7 miles) of the atmosphere, and only the lower half of that contains air sufficiently dense to breathe. With increasing height above sea level, air gets less dense and each lungful delivers less oxygen to the blood. At 5 km (3 miles) up, a lungful of air contains half the oxygen of a lungful at sea level, and people risk falling prey to acute altitude sickness, which can strike with deadly swiftness. Even so, human beings adapt. Mountaineers cope with high climbs by first spending several days at an intermediate altitude, giving their body time to manufacture more of the red blood cells that extract oxygen from the lungs. Over long periods of time, whole communities have evolved a tolerance to altitude. La Rinconada in the Peruvian Andes is probably the highest permanently inhabited town on Earth, with a population of 7000 people living at an altitude of 5.1 km (3.1 miles). The indigenous residents probably inherited their ability to tolerate altitude from generations of Andean ancestors, but even so their babies tend to be unusually small, perhaps because the low oxygen level limits the flow of nutrients to the foetus during pregnancy. When the Spanish conquistadors came to the Andes in search of silver and gold in the sixteenth century, they proved less adaptable. The Spanish built a large mining town, Potosí, some 4 km (2.5 miles) up in the Bolivian Andes. Writing in 1639, the scholar and monk Antonia de la Calancha recorded that, for 53 years, every single child born to a Spanish woman died either at birth or shortly after.

BELOW **THE INDIGENOUS PEOPLE OF THE PERUVIAN ANDES HAVE ADAPTED TO SURVIVE IN PLACES WHERE THE AIR PRESSURE CAN BE AS LOW AS HALF THAT AT SEA LEVEL.**

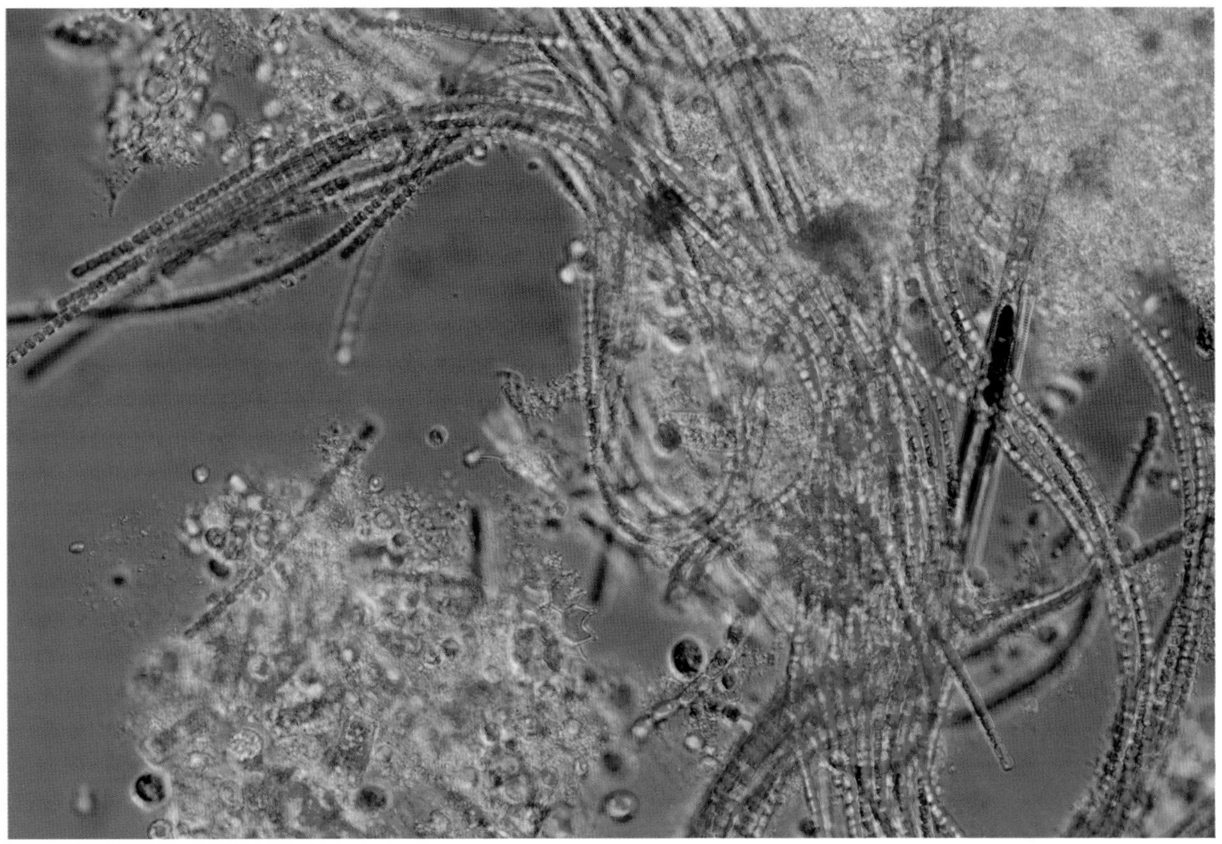

ABOVE **MICROSCOPE VIEW OF MODERN CYANOBACTERIA GROWING IN FILAMENTS. ORGANISMS MUCH LIKE THESE WERE AMONG THE EARLIEST LIFE FORMS ON EARTH.**
OPPOSITE **LIVING STROMATOLITES IN SHARK BAY, AUSTRALIA. THE CYANOBACTERIA THAT CREATED THESE STRUCTURES ARE DIRECT DESCENDANTS OF THE FIRST PHOTOSYNTHESIZERS ON THE PLANET.**

around which ships' mooring ropes are wound. But they are not stone. Touch the ones just below the surface and they feel slimy; look closely and you may be lucky to see the tiny bubbles rising from them. For they are living, or at least the top layer of them is living. The mounds are called stromatolites, from the Greek for 'mattress of rock', and they have been growing, layer upon thin layer, for millennia. They are built by microscopic organisms called cyanobacteria (sometimes misleadingly called blue-green algae, though they aren't algae) growing in 'microbial mats', each new layer forming on top of the dead remains of thousands of previous generations. It takes centuries for the mounds to build up, rising less than a millimetre per year, and when they begin to protrude from the water they stop growing. A sand barrier restricts the flow of sea water into Hamelin Pool, which has made its salinity rise to about twice that of normal sea water, and few creatures that might graze on microbes can survive in that concentration of salt. The result is that the stromatolites here have flourished.

The discovery of the Shark Bay stromatolites in the mid 1950s was quite startling to science, for stromatolites had hitherto been known only as fossils, and very old fossils at that. In fact, the world's oldest known fossils are stromatolites dating back to 3.5 billion years ago,

Blue Sky, Red Sunset

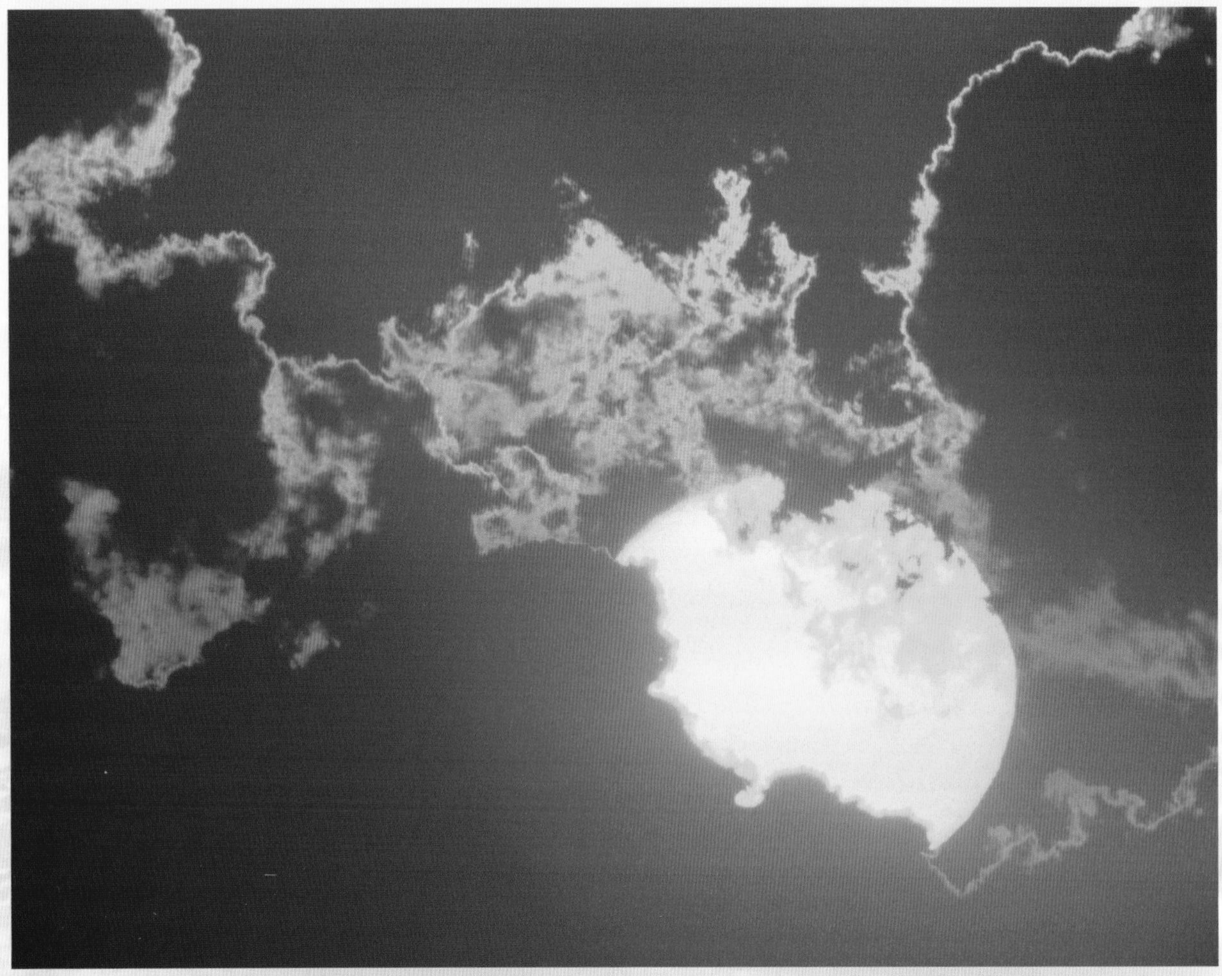

ABOVE **LIGHT FROM A LOW SUN TRAVELS THROUGH SO MUCH ATMOSPHERE THAT THE BLUE LIGHT IS SCATTERED AWAY, LEAVING A RED SUNSET.**

The Sun's visible light is made up of a number of colours, which can be spread out into the spectrum familiar to us in the rainbow. Light travels in waves, and each colour has a distinct wavelength. Or to put it another way, the waves move at a slightly different frequency in each colour. Red light, at one end of the spectrum, moves with the lowest frequency and has the longest waves. Blue, near the opposite end, moves with a higher frequency and has nearly the shortest waves. By sheer coincidence, the size of the nitrogen and oxygen atoms in air is such that they vibrate at about the same frequency as blue light. As a result, when sunlight strikes the atmosphere, the blue light literally collides with the atoms of these two gases, making them vibrate, and then bounces off in random directions. The scattered blue light, which we see coming from every direction, gives the sky its colour. By contrast, red light, with its longer wavelength, passes straight through the air. However, when the Sun is low on the horizon, its light strikes the atmosphere at a shallow angle and so must pass through much more air to reach our eyes. All the blue light gets scattered away, and only long wavelengths such as red and orange get through. Hence, at sunrise and sunset, the sky is red. Earth's primordial atmosphere had far less oxygen than today and would have scattered less blue light, so it would have been a much more yellow or red colour, an effect enhanced by all the dust still swirling around the young planet.

and many fine examples have been found in the empty lands of Western Australia, not far from their living descendants in Shark Bay. As you stand at the edge of Hamelin Pool and look out to the Pacific Ocean, you get a sense that you're looking back into the distant past and glimpsing what the world was like 3.5 billion years ago, when bacteria dominated life on Earth and built stromatolite reefs that bounded shorelines around the globe. The only difference is that the sky was orange rather than blue, for the billion-year-old Sun burned more faintly than today and the young Earth's atmosphere was laden with dust and rich in carbon dioxide.

To think that life forms can survive almost unchanged for more than 3 billion years is startling, but it is what the stromatolites did to our world that entitles them to a special place in our affections. The bubbles of gas rising from their green, slimy surfaces offer a clue. The gas is oxygen, for the stromatolites are carrying out photosynthesis – the process that enables the myriad varieties of plant life on Earth to feed on the energy of the Sun. Cyanobacteria and plants draw carbon dioxide from the air and, using the power of sunlight, combine it with water to make sugars, giving off oxygen as a waste product. And as every schoolchild learns, animals perform the reverse trick, using oxygen from the air to unlock the energy trapped in sugar molecules and emitting carbon dioxide as waste, which in turns feeds plants, continuing the cycle. It's a perfect, harmonious relationship between plant and animal and atmosphere, but it wasn't always that way. There was little or no oxygen in the air above the early Earth. The primordial atmosphere was produced by 'outgassing' of materials that had been trapped in the planet's interior during its accretion. As we saw in the previous chapter, heavy materials like iron and nickel sank into Earth's core, while lightweight compounds bubbled up to the surface. Large quantities of water were trapped in the young planet, and this would have boiled to the surface as steam, emerging through the millions of volcanoes that erupted all over the world, along with sulphurous fumes, carbon dioxide, nitrogen, methane and inert gases like neon. This was the composition of the early atmosphere, and to us it would have been pure poison.

The first liquid water probably began to condense on to the surface about 4.4 billion years ago, initially as pools and lakes. As the planet cooled they grew in size, eventually becoming oceans. Some time after that, and astonishingly quickly, life appeared. How and where life began are among the great unanswered questions of science and will probably remain so for a very long time (*see* 'First Life', page 114). The discovery in the last two decades of 'extremophiles' – bacteria that thrive in what seem like terminally hostile environments, such as scalding volcanic springs – has tilted the argument in favour of the sea floor as the birthplace of life. But wherever life began, it certainly seems to have got off to a quick start. Its tell-tale signature – a particular ratio of the isotopes carbon-12 and carbon-13 – can be detected in rocks found at Isua in Greenland that have been dated to 3.8–3.9 billion years ago, which is around the time that the late heavy bombardment came to an end and Earth ceased to be pounded by asteroids and comets.

The very first life forms were probably simple carbon-based molecules with the fluke ability to replicate themselves. Whatever they were, once they'd established themselves they began to multiply, evolve and increase in complexity, eventually giving rise to single-celled microbes much like bacteria. These microscopic bugs ruled the planet for more than 3 billion years. The earliest of them probably derived their energy from chemicals (chemosynthesis), but when the bombardment era ended and Earth's surface became a safer place to dwell, some of them switched to sunlight. They began to photosynthesize.

Proliferating in sunlit shallow seas, they grew into stromatolites. Just as plants do today, the prehistoric stromatolites absorbed carbon dioxide from the atmosphere

First Life

We do not know exactly how, when or where life arose on planet Earth, but there is no shortage of theories. For some scientists, it arrived aboard meteorites blasted off Mars or even Venus, whose early environments may have been kinder and gentler to life than Earth's. Dispatched frozen into space, the microscopic Martians or Venusians stayed dormant inside tiny rock crevices, screened from the Sun's lethal radiation, until they crash-landed on Earth. For other authorities, life was spawned in the shallows of early oceans, though it is difficult to imagine anything surviving when the planet's surface was being pelted by asteroids and irradiated by unfiltered sunlight. So for many, the most likely site for the origin of life was the relative safety and obscurity of the deep ocean floor.

The ocean depths never dried out, never got too hot or cold, and never became too acidic or alkaline. True, the first oceans were sterile waters, ultra-dilute and short of vital minerals. But seeping into them from below was a warm stew of nutrients formed where sea water percolated through hot volcanic rock and erupted back out through 'hydrothermal' vents. These hot springs carried nitrogen (as ammonia), sulphides, phosphates and tiny amounts of iron, nickel, manganese, cobalt and zinc – an ideal combination of chemical building blocks for fledgling microbes.

Today, hydrothermal vents are mostly found along the mid-ocean ridge where new crust is forming. Sometimes called 'black smokers', they spew out clouds of dark metal particles in water heated to a scorching 400°C (750°F). Some are actually more like cooling towers than chimneys, their plumes of hot water rising for 1400 metres (4500 feet), spreading to a kilometre wide and drifting down-current for tens of kilometres. But they seem an unlikely candidate for the cradle of life, given that their discharge is highly acid and hot enough to melt lead. Away from the centre of the mid-ocean ridge there are cooler but equally nourishing outpourings. In 2000, geologists discovered a spectacular cluster of such springs on the Atlantic floor.

BELOW **THE SCALDING, TOXIC WATERS OF A GEOTHERMAL POOL IN NEW ZEALAND ARE HOME TO A FLOURISHING POPULATION OF LIVING THERMOPHILES.**

Near-boiling, metal-laced alkaline water was seen streaming out of wondrous submarine spires and towers that earned the site its name: the 'Lost City'.

To see what kind of organisms might have inhabited the strange ecosystems of the early Earth, we have to visit one of the most hostile environments on land. Rotorua in New Zealand's North Island is a volcanic hotspot. Although only a couple of million years old, which is young by geological standards, it bears certain similarities to the infant Earth. It is an unworldly land of perpetual volcanic activity, with noxious gases bubbling up through steaming mudpools and lurid chemical stains decorating the ground. At Waitapu, just south of Rotorua, is the famous Champagne Pool, a multicoloured chemical cauldron. Its waters are 75°C (167°F) – hot enough to blister fingers – and are laden with foul-smelling hydrogen sulphide and lethal quantities of arsenic. And yet, literally billions and billions of organisms flourish in this toxic soup. They are 'thermophiles', heat-loving bacteria that use sulphur and other chemicals to fuel their unusual metabolism. Unlike cyanobacteria, which use sunlight to manufacture food through photosynthesis, thermophiles do it with chemicals: chemosynthesis. In the Champagne Pool they tolerate temperatures comparable to those of deep-sea vents, and they carpet the rocks by the million, forming a fur of orange fibres. The water may be deadly to humans, but to the bacteria living here the witches' brew of chemicals is a veritable Eden.

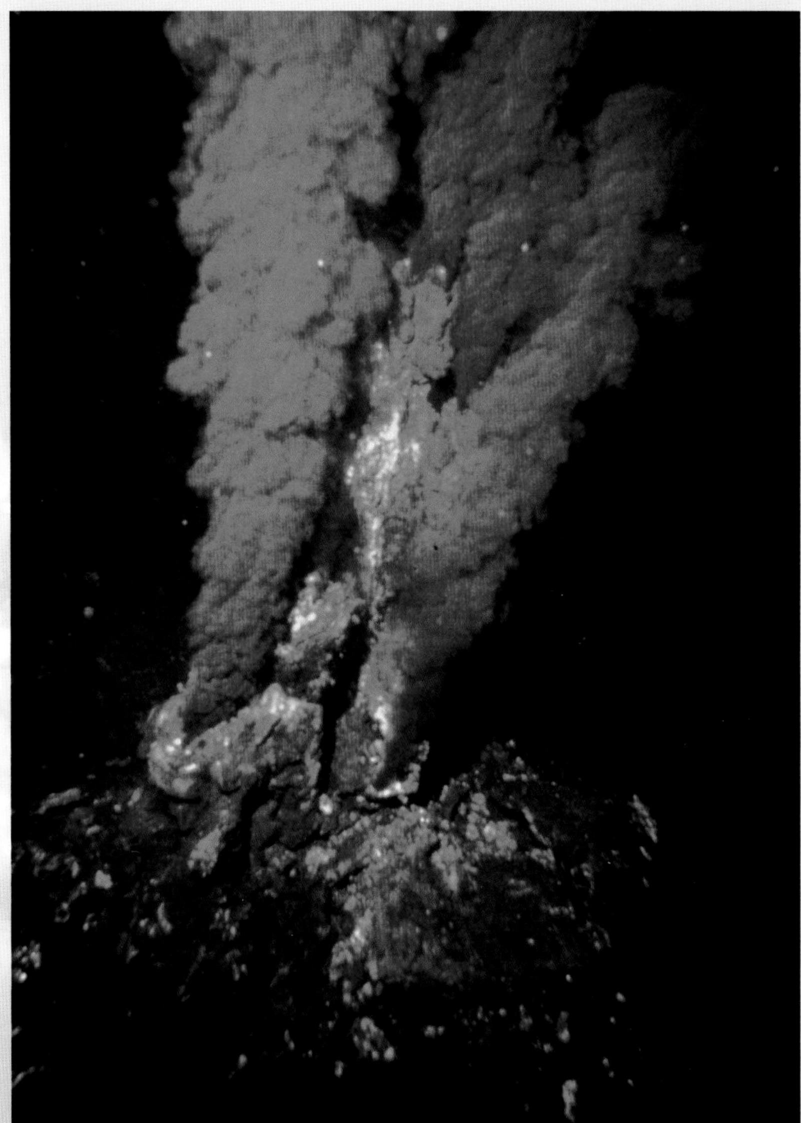

ABOVE **SCALDING, CHEMICAL-LADEN WATER GUSHES FROM A HYDROTHERMAL VENT.**

We have no idea whether the microbes that flourish here are exactly like those that kick-started life 4 billion years ago, but it is likely that the Earth's early organisms lived by chemosynthesis and could tolerate extreme environments. In fact, we find modern extremophiles in all sorts of unlikely places, from steamy volcanic wonderlands like Rotorua to the frigid wastes of Antarctica. Bacteria have been found living on rock in the deepest mines and in ocean mud hundreds of metres beneath the sea floor. Some have even been found in deep-sea sediments millions of years old, yet still alive, in a state of suspended animation. Even in boreholes thousands of metres deep that penetrate Earth's bedrock, so-called intraterrestrial life has been found alive and well. The upper part of Earth's crust is probably seething with them, forming a vast realm of life that people have begun to call the 'deep biosphere'.

and expelled oxygen bubbles as waste. And so, bubble by bubble, Earth's unbreathable early atmosphere slowly began to change. But it was to take a very long time for the stromatolites to make a big difference, because for the next billion years or so, something got in the way.

The Age of Rust

The world extracts around 1.3 billion tonnes of iron ore from the ground every year to meet the increasing demand for steel, and the figure is growing. All of it comes from vast, open-cast mines that grow like scars in the land. One of the largest is in Hamersley Basin in Western Australia, where 80 million tonnes of iron ore are excavated each year. The ground here is relentlessly drilled and blasted, and huge tractors shovel the rust-coloured ore into gigantic trucks that can carry 240 tonnes at a time. The resulting man-made canyon is 5 km (3 miles) long, its sides terraced by the winding tracks cut into layer after layer of bedrock. Viewed from an aircraft, or even from a satellite, the vast hole in the ground looks like a giant's fingerprint.

The iron is extracted from an ancient type of rock that no longer forms on Earth. The rock was laid down as sediment on the sea floor billions of years ago, and it contains a beautiful pattern that geologists call a banded iron formation. Slender stripes of reddish iron oxide – a chemical more familiar to us as rust – alternate with grey stripes of silica and shale, the parallel layers curving in sinuous undulations and whorls. Almost all the world's iron comes from banded iron formations, and they are found in some of the very oldest rocks on the planet. But what's even more remarkable is that the oxygen trapped within them to make the rust was produced by the earliest stromatolites.

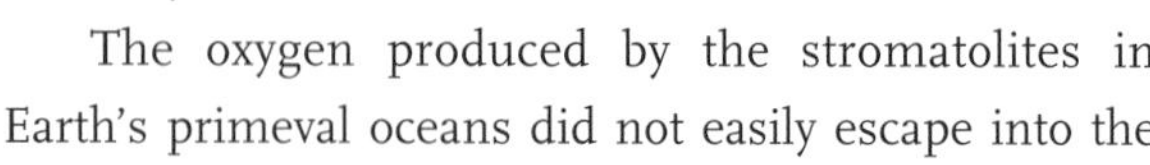

The oxygen produced by the stromatolites in Earth's primeval oceans did not easily escape into the

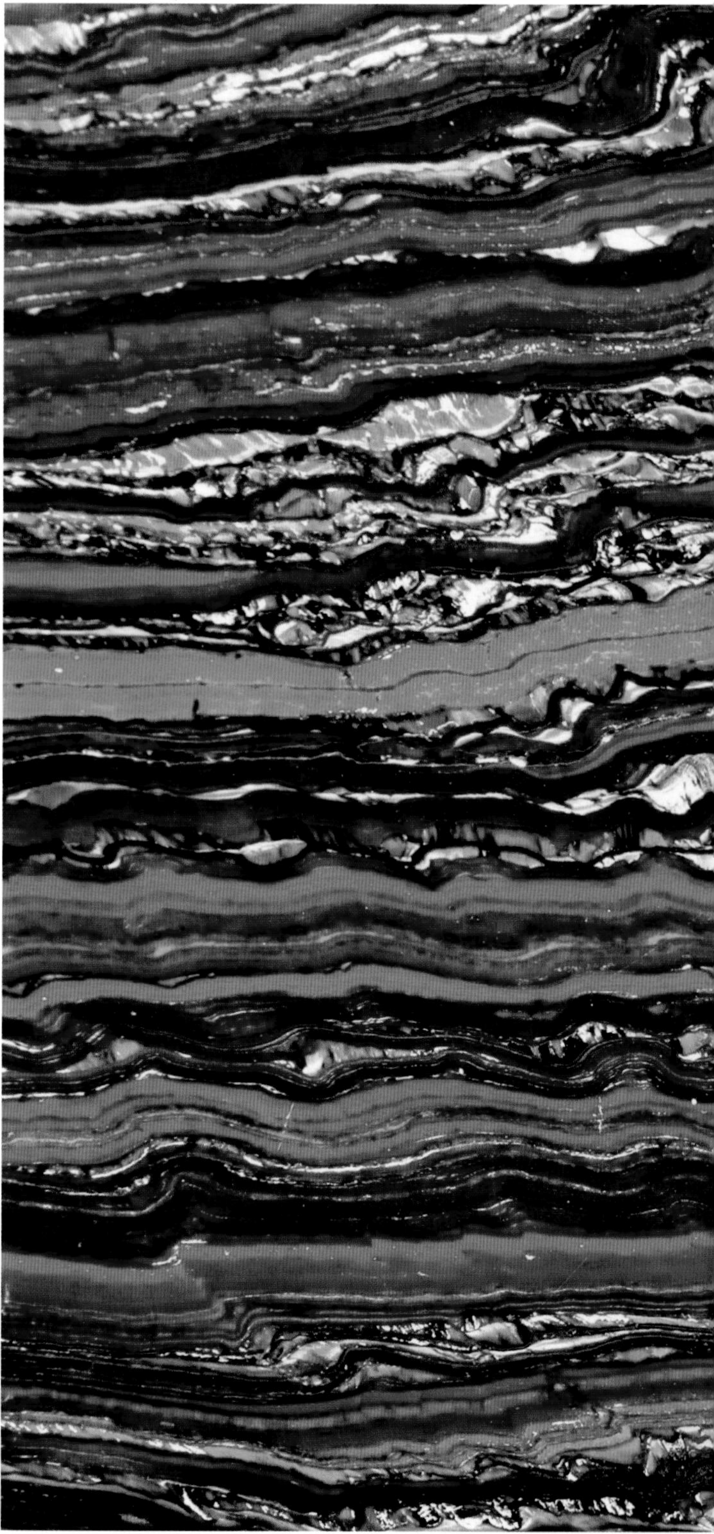

RIGHT **THE LAYERS OF RUSTY ROCK THAT MAKE UP A BANDED IRON FORMATION.**

The Ozone Hole

You might be forgiven for never having heard of Thomas Midgley, Jr. He was an American mechanical engineer and chemist who died tragically in 1944, strangled in the cables of a hoist he'd devised to lift him out of bed after becoming disabled by polio. Before his death, Midgley was awarded two prestigious medals by the American Chemical Society for his contributions to chemistry. Perhaps fortunately, he did not live to discover the disastrous impact of his inventions. For one of them was the anti-knock lead additive for petrol, now recognized as a dangerous pollutant, and the other was the refrigerator coolant 'freon', the first of the chlorofluorocarbons (CFCs). CFCs were the answer to a consumer prayer: previous refrigerator coolants were toxic and explosive, but CFCs seemed totally inert. Then they turned out to make perfect propellants for aerosol cans, and their use spread even further.

It was about 30 years after Midgley's death that the true consequences of his genius became clear. Celebrated during his life, he is now described as having had 'more impact on the atmosphere than any other single organism in Earth's history'. For CFCs were silently and invisibly wreaking havoc in the atmosphere, where they were creating an ever-greater hole in our planet's protective ozone layer.

In 1971, the British scientist James Lovelock (father of Gaia theory) joined a research cruise in the South Atlantic to study the atmosphere. His measurements led to an alarming discovery: all the CFCs that the world had ever released seemed to be lingering in the atmosphere, though at that stage Lovelock could not see what harm they might do, other than add to the haze. Then, in 1974, the US and Mexican chemists Frank Sherwood Rowland and Mario Molina found that CFCs were not quite as inert as people had thought. High in the atmosphere, they were broken down by ultraviolet light to release highly reactive chlorine atoms, which then attacked ozone. It is ironic that Molina's name now appears in the same list of chemistry award winners as Midgley's, but for having shown the terrible consequences of the earlier man's invention.

Measurements taken in Antarctica in the early 1980s shocked the world with the news that ozone was disappearing far faster than anyone had thought possible, and today a gaping ozone hole forms over the poles every winter, threatening the people below with exposure to the most harmful types of UV radiation. But there is a hopeful lesson to be learned from the story. Once the severity of the problem had been recognized, the international community moved swiftly and a worldwide ban on CFCs was imposed. The ozone hole still forms, but it has stabilized and will slowly heal over future decades. It shows that when governments awake to an environmental threat, they can do something about it – giving us hope yet for a similar response to global warming.

BELOW **A FALSE-COLOUR SATELLITE IMAGE OF THE OZONE HOLE THAT OPENS OVER ANTARCTICA EVERY WINTER.**

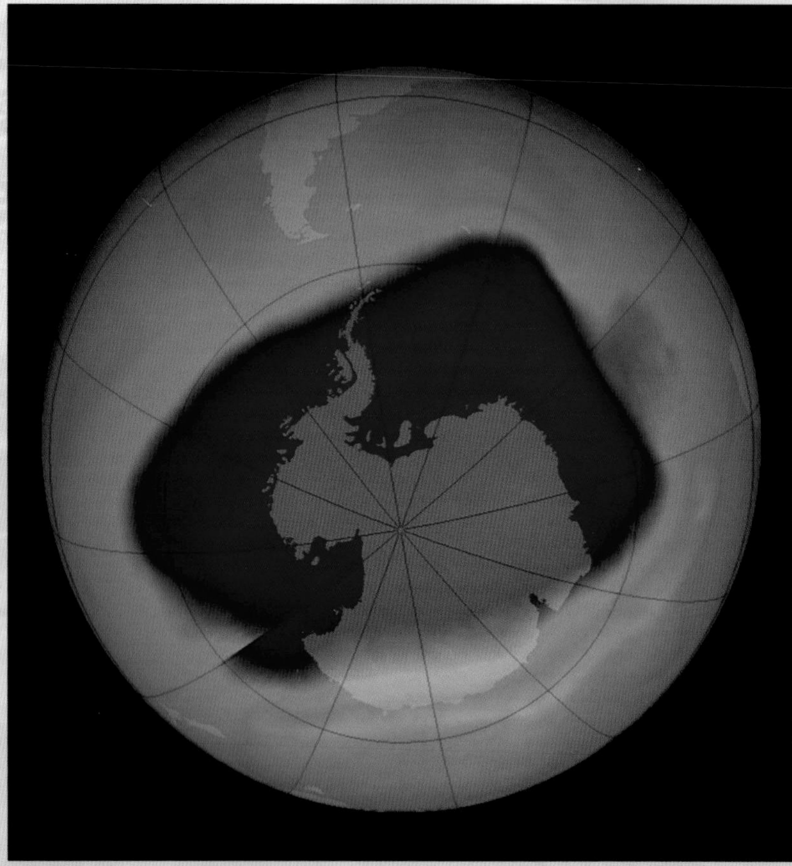

atmosphere. The early oceans were foul by our standards. Today, the Black Sea provides a clue as to what they were like. Its deepest waters contain very little oxygen and are almost stagnant, the most complex form of life being tiny nematode worms that can complete their life cycle without oxygen. Bacteria also thrive in the depths, generating hydrogen sulphide as a waste product, and from time to time large upwellings of the gas bubble to the surface, releasing foul odours. Three billion to 4 billion years ago, the oceans were just as stagnant, with little oxygen and only bacterial life. The water also contained iron salts leached out of Earth's crust, and any oxygen released by the stromatolites would have reacted with the iron to form iron oxide – rust. The rust sank into the depths of the sea to be deposited in fine layers, building up year after year, century after century, millennium after millennium, the pace dictated by the slow, layered growth of the stromatolites.

This process happened on a global scale and carried on for more than a billion years. And little by little, the iron began to get used up. Once it was all gone, the oxygen gradually saturated the sea water until it had nowhere to go but up. Then, slowly but surely, starting about 2.7 billion years ago, Earth's atmosphere was enriched with the gas we depend on today, and the sky lost its sickly reddish tinge and became a familiar blue.

It did not stop there. As the concentration of oxygen in the atmosphere rose, oxygen molecules began to get broken down by ultraviolet light in the upper atmosphere, forming individual atoms. These then recombined in triplets to become ozone, an unstable molecule that continually reverts back into normal oxygen, only for the cycle to be repeated. Thus it was that a layer of ozone formed in the stratosphere and began to screen Earth's surface from the most damaging types of ultraviolet radiation. So the stromatolites not only created the atmosphere that we have evolved to breathe, but also provided protection from the Sun's sterilizing rays, making our planet's surface a safe place to live. We owe them a lot.

Deep Freeze

Around 2.5 billion years ago, the level of oxygen in the atmosphere began to rise rapidly, marking the peak of the first great tussle between oxygen and carbon dioxide. This point in Earth's history coincides, in geological terms at least, with the first of several apocalyptic glaciations in which ice seems to have enveloped the entire planet, leading to a Snowball Earth (see chapter five). It is unclear what triggered the great freeze of 2.45 billion years ago, but one theory is that oxygen was to blame. The rising proportion of oxygen in the atmosphere would have lowered the level of other gases, in particular methane – a greenhouse gas 21 times more potent than carbon dioxide. With less methane to trap the Sun's warmth, the planet's climate cooled until it passed a 'tipping point' that allowed ice to take over. Earth was trapped in the freezer. How long the cold snap lasted is unclear, but it seems that a steady build-up of carbon dioxide from volcanoes eventually brought it to an end. Some scientists reckon the carbon dioxide level rose until there was 350 times more than today, causing a massive increase in the greenhouse effect. Carbon dioxide had fought back.

By about 2.2 billion years ago, oxygen levels had crept up again, reaching between 5 and 18 per cent of modern levels. For the next 1.5 billion years, the atmosphere seems to have settled into an eon of equilibrium, the climate kept equable by the workings of Earth's global thermostat (see chapter two). Little is thought to have changed during this calm period, apart from one thing that needs to be mentioned in passing. Around 2 billion years ago (or, according to some scientists, as early as 2.7 billion years ago), a new form of life emerged. Unlike bacteria, whose simple cells have no nuclei, the new organisms had complex and highly organized cells with their DNA packaged up in a cell nucleus. Members of this new evolutionary line got their energy by using the increasingly abundant oxygen to break down organic molecules. They were the ancestors of nearly every multicelled organism alive on Earth today – including ourselves.

The period of calm came to an end around 780 million years ago, when the global thermostat broke down and Earth was again plunged into a succession of snowball conditions, this time lasting 150 million years. Once again, greenhouse gases came to the rescue. After the thaw, life took a giant leap forward: the age of bacteria was replaced by the age of animals. Multicellular creatures large enough to be seen with the naked eye left fossilized traces of their bodies – weird and wonderful things like *Parvancorina*, a tiny, soft-bodied animal shaped like a shield, and *Dickensonia*, which looked a bit like a straggly bath mat and grew to a metre (3 feet) long. This menagerie of soft-bodied animals is known collectively as the Ediacaran fauna, after the small town of Ediacara in South Australia where their fossils were first discovered.

The Ediacarans enjoyed a quiet life, lounging on the sea bed soaking up nutrients through their skin. But soon new life forms evolved – animals equipped with rigid skeletons, gripping claws and protective armour. The soft, peaceful garden of Ediacara turned into a savage world of predator and prey. With the ice gone and the global thermostat back in operation, the evolution of life advanced at a breakneck pace. An explosion of new types of animal spread through the oceans, spawning many of the major animal groups that exist today.

Land of Fire and Giants

The explosion of new animal forms took place about 530 million years ago, and from that era onwards, the build-up of oxygen continued apace and became a driving force in evolution. Between 400 and 500 million years ago, life conquered the land, with simple plants such as mosses and liverworts leading the charge, followed swiftly by small animals such as millipedes. As oxygen levels climbed, the new ecosystems grew in stature and complexity, until lush, swampy forests had developed. And so began one of the most extraordinary periods in Earth's history: the Carboniferous, named after the deep beds of coal that formed from the remains of the trees. Those ancient forests, which are now mined all over the world, were home to some of the most remarkable creatures ever to inhabit the planet.

ABOVE AND OPPOSITE TOP **FOSSILISED PLANTS FROM THE CARBONIFEROUS ERA.** OPPOSITE BELOW **FOSSIL OF A GIANT DRAGONFLY FROM THE CARBONIFEROUS ERA. IT COULD NOT HAVE FLOWN IN TODAY'S ATMOSPHERE.**

In the coal mines of Derbyshire in England, miners often used to find imprints of leaves, which they would chisel out to take home to excited sons and daughters. But in 1979, miners working in the town of Bolsover stumbled upon a quite different fossil altogether. It was a dragonfly, but a dragonfly with a difference: its wingspan was more than half a metre (1.6 feet). Anyone with a fear of creepy-crawlies would have found the Carboniferous a truly scary place, because the Bolsover dragonfly was one many giant creatures that thrived at this time. There were scorpions 1 metre (3 feet) long, millipedes even longer, mayflies with wingspans of 40 cm (16 inches), a monstrous, spider-like creature called *Megarachne* that was half a metre (20 inches) wide, and amphibians 5 metres (16 feet) long. The plants were huge too. The lycopods, which we know today as tiny club mosses, reached heights of 50 metres (160 feet). All these life forms were giants by today's standards, so what was going on? A hint at the answer comes from the way the animals breathed. Insects do not have lungs. Instead, their bodies are riddled with tiny tubes called trachea that let air diffuse in and out passively. This puts a size constraint on insects, because the passive breathing system cannot deliver oxygen quickly to deep tissues. In today's atmosphere, a dragonfly the size of the Bolsover specimen would not be able to beat its wings fast enough to fly, because its flight muscles would be deprived of oxygen. To get airborne, it must have flown through air much richer in oxygen than ours.

Calculations suggest the Carboniferous air was 30–35 per cent oxygen. The explanation for this high level of oxygen lies in the coal itself. Coal forms when dead plant material doesn't fully decompose, instead becoming buried and then compressed over vast spans of time. By 375 million years ago, the swampy forests of the Carboniferous were already laying down the first coal seams at a prodigious rate, and they continued doing so for millions of years. In fact, about 90 per cent of the world's coal stems from this time. The Carboniferous plants must have been piling up in the ground when they died, perhaps because the swampy ground was stagnant, or perhaps because bacteria had not yet evolved the ability to digest the tough fibrous material (lignin) in wood. And because the dead plants weren't being fully recycled, there was an imbalance in the carbon and oxygen cycles. The trees were taking in carbon dioxide and giving off oxygen through the process of photosynthesis, but the reverse process wasn't happening – the carbon was staying trapped in the plants, and the oxygen wasn't being used up by the

ABOVE **THE SWAMP FORESTS OF THE CARBONIFEROUS ERA REGULATED THE COMPOSITION OF THE ATMOSPHERE, JUST AS THE RAINFORESTS DO TODAY.**

process of decay. So as forests flourished, producing more and more oxygen, the atmosphere's oxygen level rose. Life, once again, had created an atmosphere in which it could make a leap of evolution.

The Carboniferous was a green world of mosses, vines, horsetails and ferns, and vast areas of equatorial land were given over to forest swamps. But the oxygen-rich atmosphere brought something else. This is the time when charcoal first appears in the fossil record, which tells us that the atmosphere now had sufficient oxygen to support fire. From around 365 million years ago, ferocious wildfires swept across the land, and they were fires like no other, for in air that was 30 per cent oxygen even the damp swamps would have burnt. Today, studies of Carboniferous charcoal reveal not only the great intensity of the fires but the nature of the plants that were consumed by them. This, along with fossil evidence, reveals that the plants had adapted to become resistant to fire, with thick bark, high crowns and long tubers that were protected deep in the soil. Destructive as the wildfires that swept the planet must have been, they were also part of the mechanism that kept the atmosphere in balance, because they would have consumed oxygen and prevented levels from rising too high. And by burning the forests, they injected carbon dioxide back into the air.

Why Deserts Exist

ABOVE **THE MAIN DESERTS OF THE WORLD ARE CLEARLY SEEN, LYING IN TWO BANDS NORTH AND SOUTH OF THE EQUATOR.**

The equator receives stronger sunlight than anywhere else, so common sense tells us it must be the hottest place on the planet. But it isn't. In fact, the hottest places are in deserts, and most deserts are thousands of miles from the equator. The explanation for this apparent anomaly lies in the circulation of the atmosphere. When warm air rises at the equator, it carries with it huge amounts of moisture from the warm tropical seas. The air chills as it rises, and so the moisture condenses into cloud droplets, and towering rain clouds form. Out in the oceans, these clouds can spiral together and grow into hurricanes, but over the continents there isn't enough water to sustain them, and they shed their moisture as tropical rain – lots and lots of it. Now stripped of water, the rising air reaches the top of the troposphere and begins to move poleward, pushed by the endless cycle of air rising below it. At around 25–30° north or south of the equator, the dry air sinks, continuing its merry-go-round journey around the great cells of Earth's weather engine. It gets compressed as it falls and begins to warm up, at a rate of about 10°C (18°F) for every 1000 metres (3300 feet). The sinking dry air creates clear skies, bereft of rain and cloud. Unlike the wet equatorial region, the land here has no cloud to screen the Sun and no water to cool the ground by evaporation, and so the temperature soars. You only have to look at a globe or an atlas, and it all becomes startlingly obvious. Laid out before you are two yellowish-brown bands a little way north and south of the tropics, showing where almost all of the world's hot deserts lie.

Who knows what might have evolved if the prehistoric forests had continued to thrive, but 250 million years ago, at the end of the Permian Period, the great flourishing of life on Earth was nearly brought to a dramatic end. As we saw in the previous chapter, life has come perilously close to total annihilation several times in Earth's history, and the worst of these mass extinctions took place at the end of the Permian. In less than a million years, 96 per cent of marine species and 70 per cent of land species disappeared from the face of the planet. Exactly why remains a matter of debate, but many scientists suspect the atmosphere dealt the fatal blow. As we saw earlier, the Permian mass extinction happened at the same time as the massive flood basalt eruption that created the stepped hills of the Siberian Traps. The eruption would have injected vast amounts of carbon dioxide into the atmosphere, resulting in dramatic global warming and possibly a 5°C (9°F) rise in ocean temperatures. And a warming of the sea could have triggered a sudden release of methane from frozen stores of methane hydrate on the ocean floor, causing an even greater hike in global temperatures – and the resulting obliteration of life.

But not every organism was extinguished. Just as life reinvented itself and reclaimed the planet after the Snowball Earth events, so new species quickly emerged to recolonize land and sea after the Permian crisis, giving rise to the age of dinosaurs. The atmosphere settled into a new equilibrium, and for the next 250 million years, carbon dioxide levels and global temperature both steadily fell – until we began to reverse that trend. As for oxygen, during the Triassic Period that followed the Permian, it fluctuated around the 15 per cent mark, then gradually rose through the Jurassic and Cretaceous to peak at 25 per cent when the dinosaurs were in their heyday, both in terms of their size and their dominance of the land. But after the great extinction of 65 million years ago, oxygen fell back, finally to hover around the 21 per cent that we enjoy today.

For much of Earth's history, the chemistry of the atmosphere has maintained a relatively stable balance and has given the planet an equable climate, keeping it habitable. Yet, viewed over the long term, it has been in a state of continual change, its various components rising and falling, sometimes gradually, at other times with catastrophic speed. The balance of gases in which we evolved to thrive is merely one point on a graph that has been fluctuating for billions of years, its spikes and troughs still far from being wholly understood or explained. But while the nature of the atmosphere is hard to predict over the long term, over the short term it is even less so. For Earth's weather is perhaps the most capricious part of our planet's workings, riven by turmoil, capable of terrible violence, and inclined to strike at random.

Air Flow

We tend to think of the air around us as empty space, intangible compared to the fluid feel of water. But in reality air is a fluid too, and just like the ocean the atmosphere is in a continual state of circulation, never still. The global circulation of the atmosphere is essential to life because it evens out the warmth of the Sun, transferring heat from the tropics, which face the full force of the Sun's glare, to the poles, where sunlight is much weaker because it strikes at a shallow angle. If the atmosphere did not redistribute the Sun's energy this way, the equator would be 14°C (25°F) warmer and the poles 20°C (36°F) colder, making complex life all but impossible over most of Earth's surface.

The relentless movement of the atmosphere – in other words, the wind – is essentially driven by the temperature difference between the equator and poles. But the wind doesn't simply blow in a straight line from equator to pole, carrying the heat with it – it's not quite that straightforward. Let's look at what goes on in the northern hemisphere to see how it all works.

OPPOSITE **SPIRAL CLOUD FORMATIONS OVER THE CAPE VERDE ISLANDS REVEAL THE FLUID NATURE OF THE ATMOSPHERE. (NORTH IS AT THE LEFT.)**

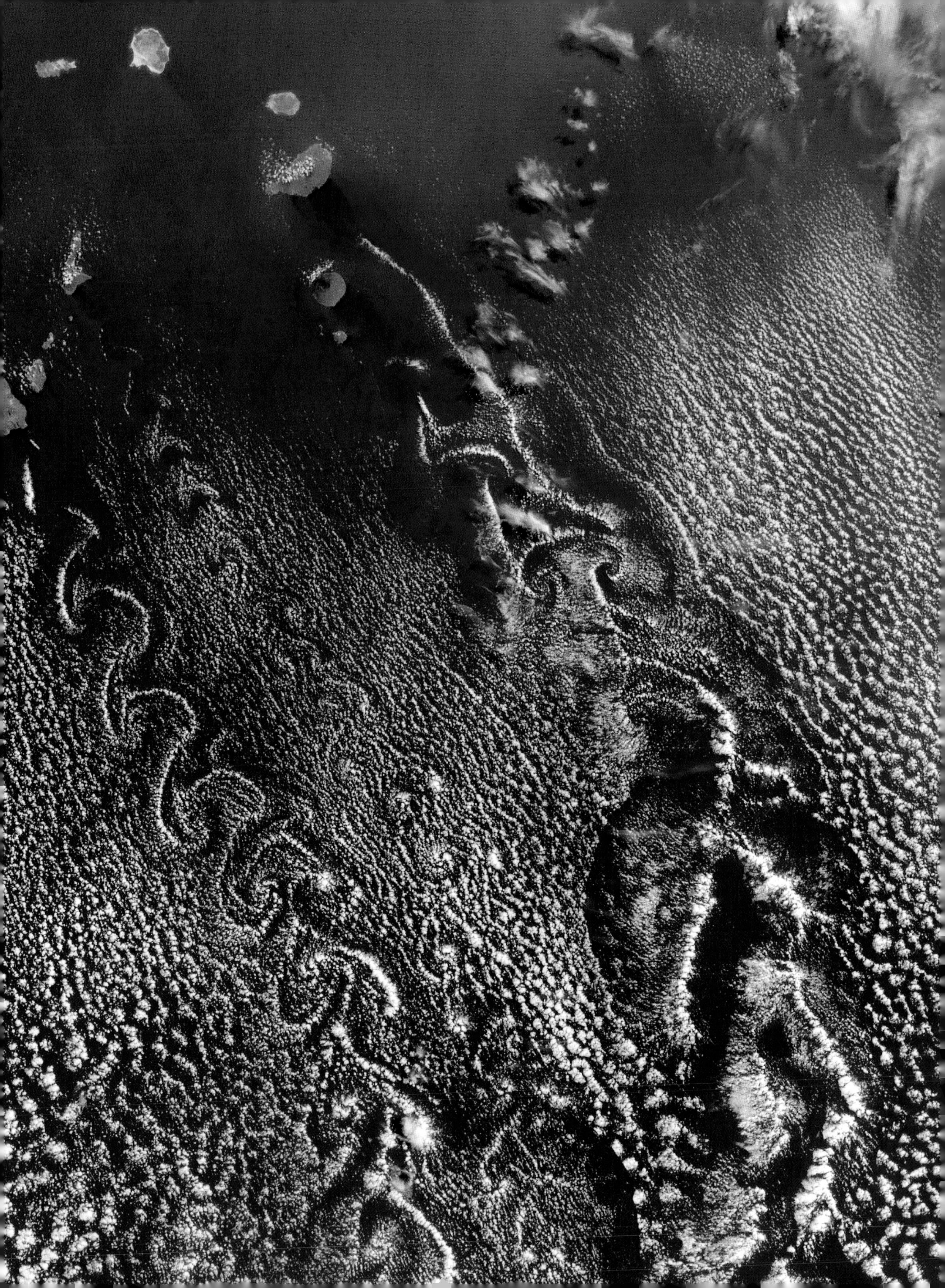

At the equator, air warmed by the Sun rises, chilling as it goes up, until it reaches the boundary with the stratosphere. It can't rise any further because it hits an 'inversion' – a point where the air temperature stops falling and begins to rise. So the mass of rising air moves horizontally instead, drifting north. It travels about a third of the way to the pole but then sinks back down, having been squeezed by the narrowing of the globe and so made denser and heavier. Then it flows back along the surface, some of it travelling south to the equator to complete the cycle, the rest going north. This circulatory system is called a cell, and there are three of them on the way to the pole. To get from equator to pole, then, the moving air must weave its way around all of them, alternately rising and sinking.

The bottom parts of the cells, where winds blow across the surface, create the predictable trade winds and westerlies that have filled explorers' sails for centuries. If Earth did not rotate, these winds would blow due north and south. As it is, they are deflected sideways by the planet's spin, making them curve to the east or west.

This inexorable flow of heat through the atmosphere is the source of wind, rain, lightning, thunder and every other conceivable weather phenomenon that bears on the lives of Earth's inhabitants. Some of our weather phenomena have been observed on other planets – there is lightning on Jupiter and there are dust storms on Mars, for instance. But in some important respects, Earth is unique. It is a curious property of our planet that it lies at precisely the right distance from the Sun for water to exist simultaneously in all three of its physical states: vapour, liquid and solid. Alone among the planets, Earth has an average temperature

RIGHT **THE GLOBAL CIRCULATION OF THE ATMOSPHERE PRODUCES AN ASTONISHING VARIETY OF STORM CLOUD FORMATIONS.** OVERLEAF **THE 'MORNING GLORY' CLOUD THAT ROLLS INTO AUSTRALIA'S GULF OF CARPENTARIA EVERY SPRING.**

ABOVE **HURRICANE FLOYD APPROACHES FLORIDA IN SEPTEMBER 1999, ITS SPIRAL CLOUD FORMATIONS AND CENTRAL EYE CLEARLY VISIBLE IN THIS SATELLITE IMAGE.**

and pressure that are very close to what scientists call the 'triple point' of water. It is water's endless transmutation from one form to another, as heat flows into and out of it, that generates much of our weather, giving us snow, rain, clouds and storms of deadly power.

And the water in our atmosphere can express itself in some truly spectacular and awe-inspiring ways. Every spring in the Gulf of Carpentaria – the vast U-shaped bay in northern Australia – an astonishing cloud formation sweeps in from the sea. It comes shortly after dawn, appearing as an extraordinarily long and straight cloud that appears to be rolling towards you. The cloud base is about 300 metres (1000 feet) high, the top perhaps 2500 metres (8000 feet), and the whole cloud can measure 1000 km (620 miles) wide. It is called the Morning Glory, and it rushes towards the land at phenomenal speed, reaching 60 km/h (40 mph). This is how a correspondent for the Sydney magazine *The Bulletin* described it in 1934: 'A low bank of clouds lined the horizon early in the morning, and gathered speed at an alarming rate. I felt sure we were going to have a deluge. On it came, a threatening, dark rolling cloud; soon the sky was completely overcast. A few drops of rain fell; then we had a delightful breeze which lasted for a couple of minutes. Away went the dark cloud as quickly as it had come, and the Sun continued to blaze as mercilessly as ever.'

Impressive as it looks from the ground, it is from the air or from space that this amazing spectacle is best appreciated. From high above, it appears like a gigantic roll of surf about to break on the shore, and since the late 1980s it has become a playground for aptly named 'sky surfers' who ride the cloud in gliders, borne aloft by powerful updraughts for as much as three hours.

ABOVE **CLOUDS MARK THE LINE OF THE INTERTROPICAL CONVERGENCE ZONE – THE BELT OF WARM, RISING AIR AROUND THE EQUATOR WITHIN WHICH HURRICANES ARE BORN.**

HURRICANE

The age of satellite imagery has afforded us myriad opportunities to watch the atmosphere in action. Some of the most striking images show clouds swirling in eddies around the tips of mountains on islands like Cape Verde, the Canaries or Hawaii. But perhaps the most compelling images are those that reveal a hurricane growing from its roots off the coast of West Africa and swelling into a vast spiral as it crosses the Atlantic and heads towards a violent end on the US coast. The way in which hurricanes form and develop illustrates beautifully how land, sea and atmosphere, underpinned by Earth's rotation, are interrelated, all acting as parts of a single complex system.

The equator lies between two of the global cells through which Earth's atmosphere circulates. The strong trade winds that flow through the bottom of these two cells converge in the middle, creating an area of weak wind known to sailors as the doldrums and to scientists as the intertropical convergence zone. In the eastern Atlantic, the converging winds arrive laden with moisture after crossing tropical waters, and where they meet they are pushed together and forced to rise, the moisture condensing into cloud. When water condenses it releases the 'latent heat' that had turned it into vapour. This extra heat makes the warm air rise faster, feeding the system and so building tall thunderclouds. More warm, humid air is drawn in at the base, and so the cycle intensifies. Tropical thunderstorms peter out quickly over land, but at sea they grow larger and can merge, forming a self-sustaining mass. This is called a tropical disturbance, and between June and November each year about 90 of them form in the Atlantic, one every couple of days. The key then is the wind above the storm. If the wind is turbulent, the storm breaks up and disappears, but if the wind is uniform,

Inside a Hurricane

A hurricane is defined as an intense storm of tropical origin, with sustained winds exceeding 120 km/h (74 mph) – and today we know more than ever about what goes on inside one. Satellite images that capture the beautiful Catherine-wheel shapes of hurricanes make them seem almost benign, like fluffy white screensavers cartwheeling gracefully across the globe. But the pretty shape belies the colossal size and power of the average hurricane, which reaches 550 km (340 miles) wide – approximately the length of Scotland – and unleashes as much energy as a 10-megatonne nuclear bomb exploding every 20 minutes.

The spiralling clouds that blow towards the centre of the storm are called spiral rain bands. They twist around the eye like the arms of a galaxy, their winds always flowing in the same direction: anticlockwise in the northern hemisphere, clockwise in the south. The warm, moist, tropical air flows ever inwards, where it builds up in a circle and rises, the moisture condensing out to create a ring of intense thunderstorms packed tightly around the central eye and towering up to 15 km (9 miles) high. This is the 'eye wall', the most destructive part of the hurricane, where the screaming winds have been known to reach 300 km/h (190 mph). At the top of the thunderstorms that make up the eye wall, the air has lost most of its moisture and flows outwards, high above the maelstrom below, cooling as it goes. Several hundred kilometres away, the cool, dry air gently sinks, bringing with it clear skies – a deceptive precursor to the storm that may be approaching.

At the centre of the hurricane, the violent thunderstorms of the eye wall release vast quantities of heat, warming the air, and so creating high pressure in the zone of cool, dry air above the storm. The pressure forces dry air down into the eye, creating bizarrely calm weather, with light winds and patches of blue sky. If you could ride inside the hurricane, you would see whirling rain clouds, with ice forming at the top, torrential rain pouring below, and lightning crackling all around you. Reconnaissance flights experience tumultuous extremes passing through the clouds, but once in the eye it is possible to take a coffee break and radio back weather reports. Pilots have even come across flocks of birds that became trapped in the eye as the hurricane formed. Unable to continue along their migratory route, they are dragged across the ocean in what amounts to a huge natural cage, until the cyclone finally fades and releases them.

BELOW **THE COST OF HURRICANE DEVASTATION WILL RISE AS CLIMATE CHANGE BRINGS INTENSIFIED STORMS.** OPPOSITE **THE EYE WALL OF A HURRICANE CARRIES THE MOST VIOLENT WINDS IN THE STORM.**

Sea of Sand

ABOVE **THE LARGEST DUNES OF A SAND SEA CAN BE THOUSANDS OF YEARS OLD. FOR SCALE, NOTE THE TINY SILHOUETTE OF A VEHICLE IN THE FOREGROUND.**

To climb to the top of a large sand dune is an exhausting process: the sand slides away beneath your feet, it burns your skin with reflected heat, and the steepness of its slope drains all your strength. But to look out over the top makes the climb worthwhile, for the view across a sand sea is breathtaking.

Sand dunes can take many different forms, reflecting the type of sand from which they are made and the pattern of the winds that shape them. A constant wind and a plentiful supply of sand can build fantastically long dunes at right angles to the wind's direction; one such dune in Mauritania is more than 100 km (62 miles) long. It takes a wind speed of about 15 km/h (9 mph) to put sand in motion, and when the speed drops below this the wind's cargo of airborne particles gets dumped.

The simplest dunes form in the lee of a boulder, a shrub, or a hill, where air has been forced to slow down behind the obstacle. Wind moves quickly over flat or rocky ground but is slowed by friction when it crosses sand – and so sand tends to attract more sand, which then builds into dunes. Their rising profiles intercept the wind, attracting yet more sand, which settles on the leeward side where the air is relatively still.

But the wind does not just add sand to dunes, it also blows it off. Fine grains get whipped up into the air, but heavier particles move along in hops and leaps, repeatedly rising and falling in a process known as saltation. On the windward side of a sand dune, the hopping sand grains climb steadily uphill, pushed by the air. If the wind is strong enough, the sand piles up ever steeper until the mound collapses under its weight. The collapsing sand comes to rest when it reaches just the right steepness – known as the angle of repose – to keep the dune stable. In the case of desert sand, this angle happens to be 30–34°.

If a dune grows taller than its neighbours, the ridge gets exposed to faster winds, and the sand is blown off, only to tumble back to the ground behind the dune. The wind soon sweeps up this loose material into the beginnings of a new dune, and so the sand sea spreads.

moist air continues to rise unchecked, and everything builds.

But something else is needed before the disturbance reaches hurricane proportions: Earth's spin. Because of the planet's rotation, surface winds bend to the right in the northern hemisphere and to the left in the southern hemisphere. The further north or south of the equator you go, the stronger this 'Coriolis effect' becomes. In summer the intertropical convergence zone drifts north, and when it reaches a latitude of 5° the Coriolis effect is strong enough to make groups of thunderstorms begin to rotate. The clouds are stirred round into organized bands, forming a spiral structure wrapped around a central point of low pressure, with the winds getting faster and faster towards the middle. The storm now takes on the look and feel of a hurricane.

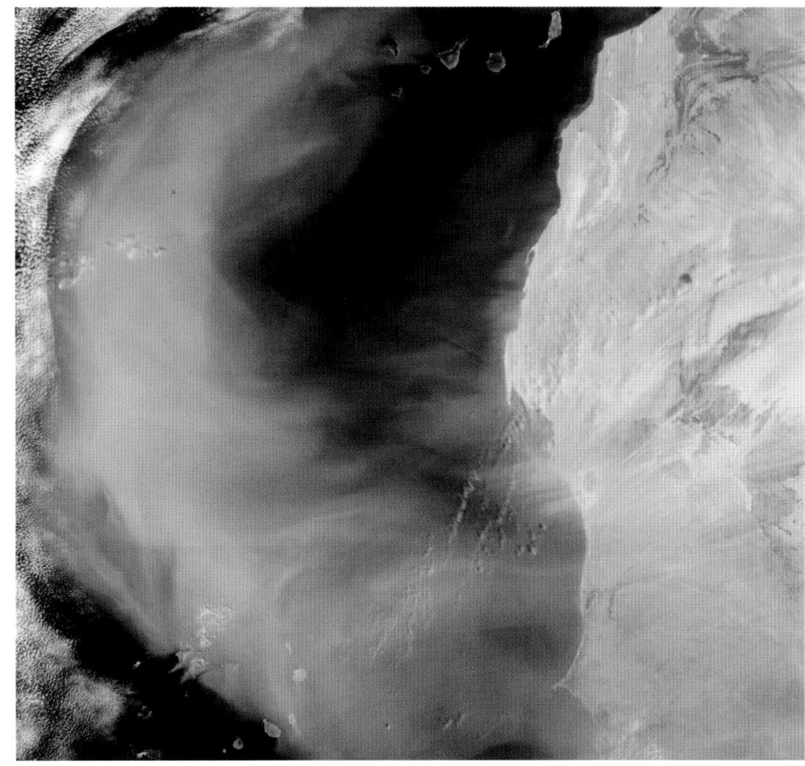

ABOVE **SATELLITE IMAGE OF A VAST CLOUD OF SAHARAN DUST BLOWING WESTWARDS ACROSS THE CAPE VERDE ISLANDS IN THE ATLANTIC OCEAN.**

A fully fledged hurricane is a vast, self-sustaining engine a hundred times larger than a thunderstorm and a thousand times more powerful than a tornado. An ordinary summer thunderstorm may have the power of three nuclear bombs; a hurricane has 25,000 times that energy, and if it stays over warm ocean water it can be fuelled for days. But the Coriolis effect that helps create hurricanes is also responsible for their inevitable demise. Because of Earth's spin, hurricanes generally drift west across the Atlantic, towards land. There they simply run out of fuel, starved of the warm ocean water that fed their growth. So the winds subside, and the hurricane downgrades to a tropical storm. It may travel on for weeks as a mid-latitude depression, before finally all trace of it is lost.

Blown Away

The atmosphere has extraordinary physical power, constantly moulding and remoulding the surface of the Earth, and nowhere is this more clear than in the vast sand seas of the world's deserts. The Sahara's great eastern sand sea – the Eastern Erg – was probably laid down over some 10,000 years and consists of dunes ranging from a few metres to more than 300 metres (1000 feet) in height. The creation of sand from rock and gravel can be witnessed on the northwestern edge of the desert where the wind whistles down Morocco's Atlas Mountains to scour the desert floor, blowing debris out towards Algeria's desert heartland. Every year, the wind brings another 6 million tonnes of sand to the vast basin in which the Eastern Erg formed.

Yet even the harsh and scouring winds of the desert play an important role in making Earth hospitable to life. On the southern edge of the Sahara is a vast, dusty basin called the Bodélé Depression. Until a few thousand years ago it was the site of a gigantic lake, but when the Sahara dried out after the ice age, the lake dwindled until only a fragment remained – Lake Chad. The rich sediments of the lake bed, scorched by the hot sun, turned into a fine dust that is easily whipped into the air. Bodélé now has the dubious distinction of being the single greatest source of airborne dust on Earth. Every four days or so, desert winds lift vast quantities of the dust high into the upper atmosphere, from where a staggering 240 million tonnes are carried out over the Atlantic Ocean each year. The bulk of it falls into the sea, nourishing plankton, but about 50 million tonnes ends up falling on the Amazon rainforest, where it provides a welcome supply of nutrients to plants growing in the region's impoverished soil. It has been estimated that more than half the Amazon basin's nutrient supply comes from the Bodélé Depression, a relationship that is a beautiful example of the planet operating as a complete biological system.

While wind is the main agent of erosion in deserts, in other parts of the world rain plays a far more destructive role. As we saw in chapter two, it is the combination of rain and carbon dioxide from the atmosphere that brings about the process of weathering, reducing even the toughest types of rock to sand and clay with the passage of time. And it is this process of weathering, brought about by the atmosphere, that keeps the carbon cycle going and our planet's global thermostat in operation, maintaining a climate perfect for life. Whatever damage we may do to Earth's climate, it is perhaps reassuring to reflect that the atmosphere will eventually correct itself and bring the planet back into balance – though it will probably take millions of years, and we may not be around to say thank you.

OPPOSITE **'THE WAVE' IN ARIZONA: A REMARKABLE EROSION FEATURE CARVED BY THE WIND OUT OF SANDSTONE ROCK THAT WAS LAID DOWN AT THE TIME OF THE JURASSIC.**

CHAPTER FOUR

OCEAN

THE WORLD OF THE ANCIENT GREEKS WAS CENTRED ON THE LANDS OF THE Mediterranean, and although they ventured further afield – into Asia Minor, western Europe, and down the Nile – their knowledge of the world was bounded by its coast. The Greek sailors who navigated as far as the Strait of Gibraltar knew of a strong current that swept past the great rock, and they decided this must be the outflow of a huge river that they could see fading to the west (the Atlantic). They called this great river *Okeanos*, and they believed that it encircled the world. It happens that the ancient Greek mathematician Pythagoras was perhaps the first person to realize the world is a sphere, but this idea was too outlandish for most of his fellow citizens, who were fairly certain they lived on the flat top of a short cylinder. Either way, they could have had no notion of the scale of what really lay beyond the treacherous currents at the gates of the Mediterranean.

By convention we divide the seas into five different oceans – Pacific, Atlantic, Indian, Arctic and Southern – but in reality they are all linked to form one body of water, more correctly termed the 'world ocean'. And as we shall discover later, the flow of water around the planet operates as a single system, so the ancient Greeks were right in principle: *Okeanos* does indeed encircle our world.

OPPOSITE **A WAVE BREAKING ON THE SHORE OF HAWAII RELEASES ENERGY THAT IT MAY HAVE CARRIED MORE THAN A THIRD OF THE WAY AROUND THE GLOBE.**

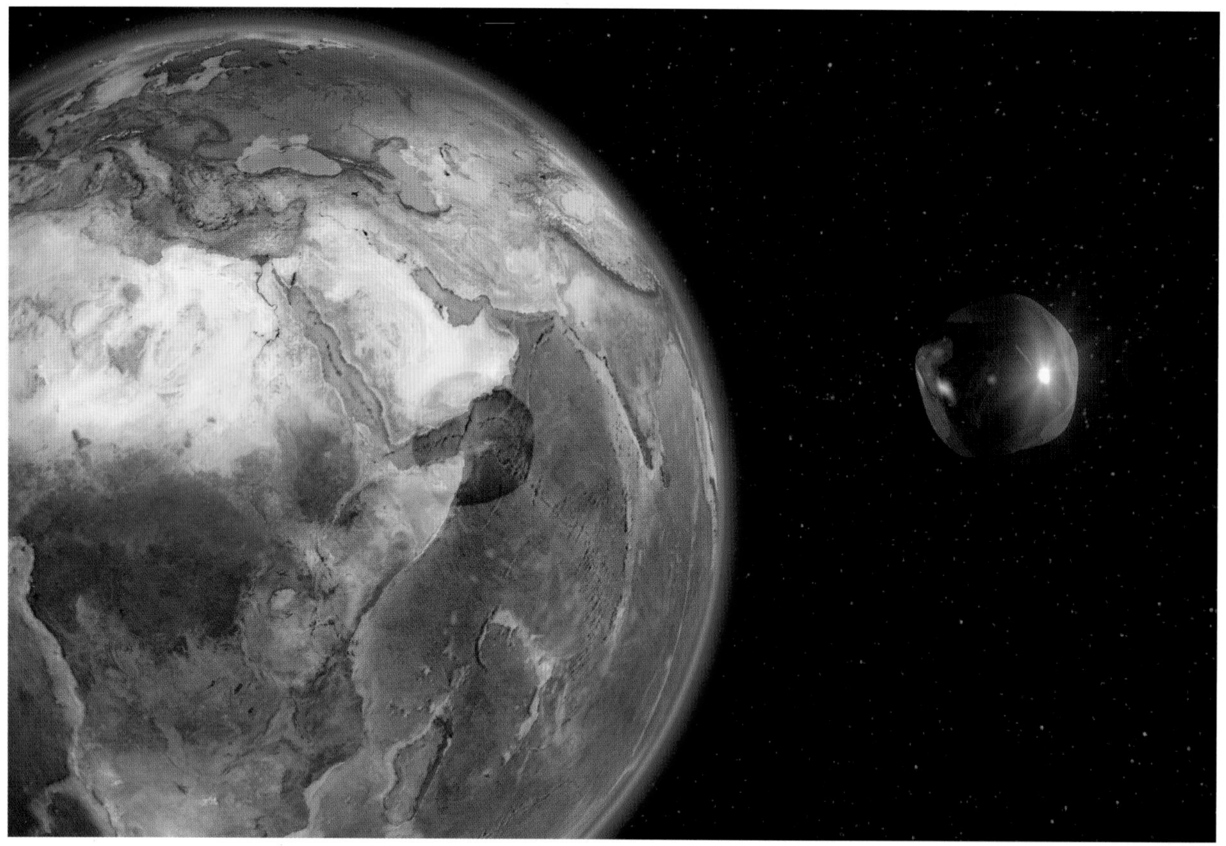

ABOVE **ALL THE WORLD'S WATER, SHOWN GATHERED INTO ONE SINGLE SPHERE, IS A SURPRISINGLY SMALL QUANTITY – YET WE DEPEND ON IT FOR LIFE.**
OPPOSITE **A VIEW OF THE PACIFIC OCEAN REVEALS HOW OCEAN DOMINATES THE SURFACE OF EARTH.**

The vital statistics of this global sea are awesome. It holds 1374 million billion tonnes of water, covers an area of 362 million square km (140 million square miles), and has an average depth of more than 3.5 km (2.2 miles), the deepest point being just short of 11 km (7 miles) down. Viewed from space, high above the Pacific, the planet appears to be a water world. You could fit all its land into the area of the Pacific and still have room for another Australia. But the view is deceptive, because ocean makes up only 0.02 per cent of Earth's mass. In geological terms, it is spread very thinly across the surface of the planet. And yet that tiny, watery fraction has been critical to the planet's evolution and also to our own existence.

Earth began to coalesce into something that resembled a planet around 4.5 billion years ago, although it effectively had to form itself all over again after the giant impact that created the Moon. The nascent planet was blisteringly hot, with oceans of molten rock but no water in sight. Things very quickly cooled down though, and as the red-hot magma solidified to form the crust and mantle, the hot rock degassed and gave Earth an atmosphere. At first the air was a scalding mixture of carbon dioxide, steam and other choking volcanic fumes, but eventually – when the planet was somewhere between 10 million and 100 million years old – things cooled sufficiently for the steam to condense and turn into rain. And so began a deluge of

ABOVE **AN IMAGE FROM *DEEP IMPACT*, TAKEN AS IT APPROACHED COMET TEMPEL 1 IN 2005.**

extraordinary proportions. Not for forty days and forty nights did the rain fall, not for months or years, nor centuries, nor even millennia. Instead, the heavens opened and water poured from the sky for perhaps as long as a million years. The early oceans thus began to form.

But that was how only part of Earth's liquid water came to be here, for calculations suggest there wasn't enough water in the planet's interior to fill the oceans we have today. There had to be another source.

All through the early period of Earth's history, the bombardment of the planet by asteroids and comets was relentless. What has become clear in the last two decades or so is that the comets brought with them water. In 2005, an extraordinary NASA space mission ended in a spectacular cosmic impact with a comet. *Deep Impact* was the name given to the probe whose task was to rendezvous with the comet Tempel 1 when it was 130 million km (80 million miles) from Earth. The probe took photographs and measurements, but its ultimate objective was to crash into the surface in a high-speed collision that was monitored by telescopes all over the world (and even by another cometary probe, *Rosetta*, which is still on the way to its own rendezvous, scheduled for 2012). The results were dramatic: over the five days that followed the impact, a vast plume of ice and dust spewed out from the comet. Thanks to the mission, scientists will now be able to calculate just how much ice comets like Tempel 1 contain. Even a small comet measuring only 10 km (6 miles) wide contains a vast amount of water, perhaps a hundred billion tonnes, and it has been calculated that a million average-sized comets colliding with the early Earth could have filled the oceans. However, it seems unlikely that comets did the job on their own, for there are subtle differences in the chemistry of sea water and comet water. But they may have provided as much as half of the water in Earth's oceans, the rest coming from inside the planet. Ironically, large comet impacts must have set back the process of ocean formation, because the vast explosions would have vaporized great swathes of Earth's surface, creating a boiling atmosphere of steam that lasted hundreds of years before it cooled sufficiently for rain to fall again. So for the early oceans, it was a case of two steps forward, one step back for many millions of years.

But form the oceans did, and life swiftly followed. As we saw in the previous chapter, life may have begun around hydrothermal vents on the ocean floor, where strange ecosystems continue to flourish even today, their energy derived from chemicals in the hot volcanic water. It is a mark of how little we really know about the oceans that these fascinating communities, which transformed our theories about the origin of life, were discovered as late as 1977. For the deep sea is still a largely unexplored realm, far too hostile for humans to venture into without the most sophisticated technology. Indeed, to date, only two people have visited the deepest point on the planet – the Mariana Trench in the Pacific Ocean – whereas twelve men have walked on the Moon.

The Deep

The exploration of the deep ocean did not begin in earnest until the 1950s, when submarine capsules able to withstand immense pressures were invented. One of the early pioneers was the Swiss scientist and inventor Auguste Piccard – the inspiration for the eccentric Professor Calculus in Hergé's Tintin books. Piccard became renowned for remarkable feats of human endeavour in the 1930s, when he designed a pressurized cabin for a high-altitude balloon and took it to a record-breaking altitude of 23,000 metres (75,000 feet) in order

BELOW **THE SWISS SCIENTIST AUGUSTE PICCARD (RIGHT) STANDING BEFORE THE HIGH-PRESSURE CAPSULE HE DESIGNED. IT BROKE RECORDS FOR BOTH ATMOSPHERIC ALTITUDE AND OCEAN DEPTH.**

to investigate the chemistry of the atmosphere and to measure cosmic rays. He then realized that the reinforced cabin would be able to withstand the immense pressures of the deep sea, and so he set about adapting it to create the 'bathyscaphe' – a tiny submarine with room for only two occupants. The bathyscaphe's moment of glory did not come until 1960, when Piccard's son Jacques, accompanied by US Navy lieutenant Don Walsh, made a daring descent into Challenger Deep, the deepest part of the Mariana Trench in the western Pacific. This vast undersea chasm marks the boundary between two of the plates in Earth's crust, though this wasn't known at the time as the theory of plate tectonics had yet to be conceived; the Mariana Trench was simply a black and impenetrable mystery.

Walsh and Piccard's descent took almost five hours, and halfway down they experienced a terrifying shock. A loud bang rang through the bathyscaphe, and they discovered that its one window had cracked. Remarkably, after checking their instruments, they reached the conclusion that the structure was still strong enough to continue, and on they went. They made it all the way to the bottom: an astonishing depth of 10.9 km (6.8 miles), putting them a full 2000 metres (6500 feet) further from sea level than they would have been on the summit of Mount Everest. They spent 20 minutes at the bottom, eating chocolate to sustain their energy and peering through the window at the muddy sea floor just below them, where they spotted a few fish – proving that the deep sea was not lifeless. It is a testament to their bravery, and to the enormity of the challenge of exploring the deep, that while thousands of people have scaled Everest to stand on the roof of the world, no-one has been back to the deepest part of the ocean.

Today, the myriad Earth-orbiting satellites give us a complete view of the ocean floor, and what stands out above all else is the extraordinary network of undersea mountains – the mid-ocean ridges. As we saw in chapter two, these run like scars across each of the five oceans, together making up a continuous mountain range 60,000 km (37,000 miles) long. It is almost entirely hidden from view, with only a handful of islands – such as the Azores, Bermuda, Ascension, Tristan da Cunha and Iceland – offering tantalizing glimpses of its peaks. The mountains stand at plate boundaries where molten rock is oozing up from below to form new ocean floor, forcing plates apart in the process (*see the illustration on page 74*). In the Afar Depression of Ethiopia, we can see this process happening before our eyes, for the valley is a new sea floor in the making. The storm of earthquakes that shook Afar in September 2005 tore open gaping fissures hundreds of metres long – local herdsmen lost goats and camels into them – and the valley floor sank by up to 100 metres (330 feet). It is already well below sea level, and at some point in the next million years or so, the Red Sea will breach its defences and rush in. But before then, perhaps 650,000 years from now, something else will happen in the region: the African plate will have inched northwards at its snail's pace of 2.15 cm (0.8 inches) a year to meet Gibraltar, closing the Mediterranean Sea. The incoming current that the Greeks had thought came from *Okeanos* will shut down, and when that happens, the Mediterranean will evaporate in a few thousand years – just as it has done in the past (*see* 'The Vanishing Sea', page 146).

Wave Power

The tectonic forces that shape the oceans over millions of years sometimes express themselves on a much shorter timescale. The earthquake that occurred off the coast of Indonesia on 26 December 2004 registered 9.2 on the Richter scale and continued for almost ten minutes, making it the longest and second most powerful on record. Deep under the sea, two tectonic plates had

OPPOSITE **WAVES, LIKE THESE SWEEPING ACROSS THE BAY OF BENGAL, TRANSPORT ENERGY ACROSS OCEANS. SEEN FROM SPACE, THE INTERFERENCE PATTERNS THAT RESULT FROM THEM COLLIDING WITH EACH OTHER ARE CLEARLY VISIBLE.**

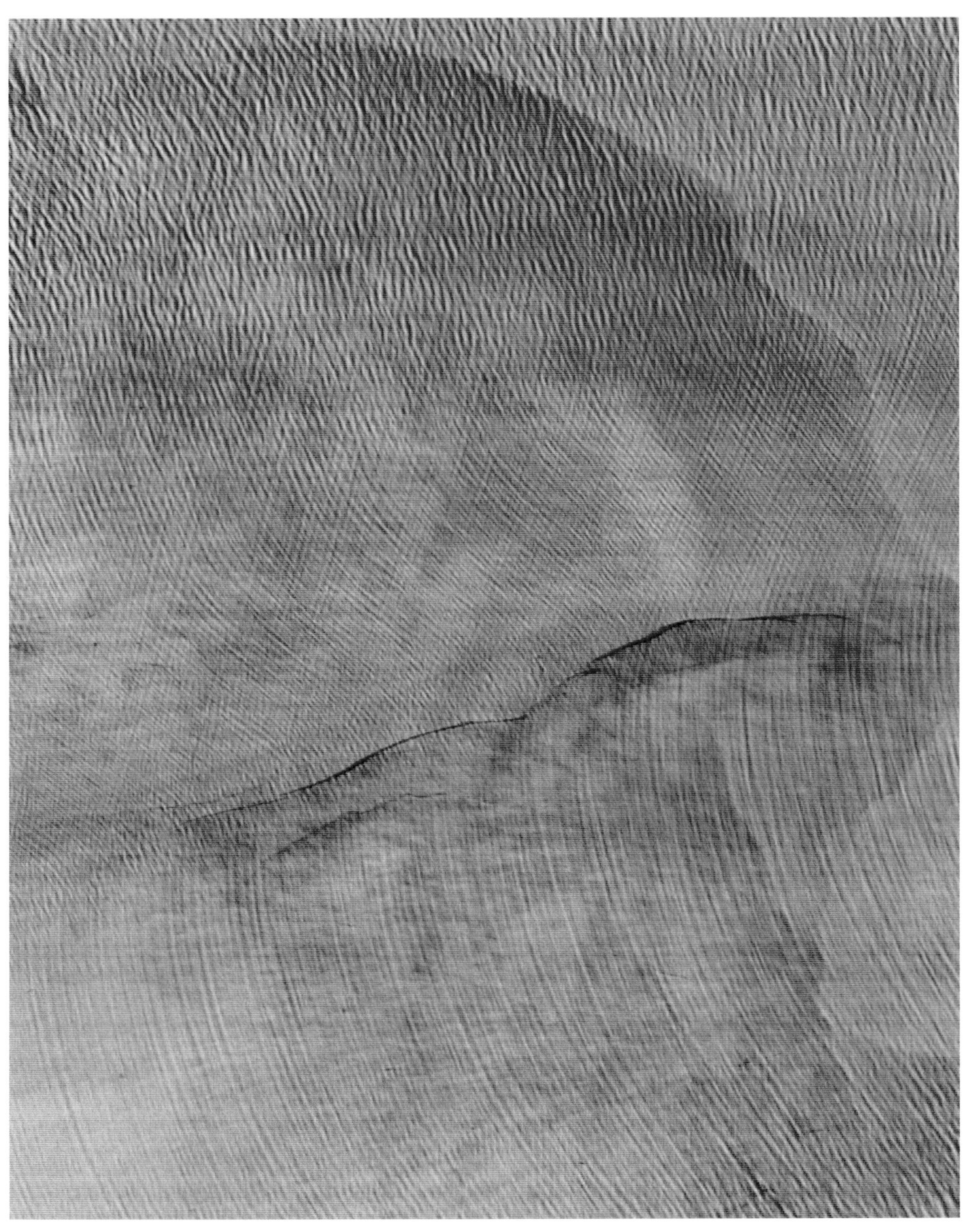

The Vanishing Sea

To us the oceans seem vast and permanent, but on a planetary timescale they are more like fleeting puddles. A little less than 6 million years ago, the Strait of Gibraltar closed and the Mediterranean was cut off from the Atlantic, triggering what geologists call the Messinian salinity crisis. Put simply, the Med began to evaporate. Enclosed seas such as the Caspian are not destined to dry out, but the closing of the Mediterranean had the bad timing to coincide with a climatic shift that pushed northern Africa's arid belt a little further north. More water evaporated from the sea than flowed in through rivers, and so the water level gradually fell by as much as 2–3 km (1–2 miles) over 500,000 years. As the shore receded, rivers such as the Nile extended their reach and carved vast canyons into the sides of the Mediterranean basin – one such canyon, now buried deep below Cairo, was longer and deeper than the Grand Canyon. The rivers continued to bring mineral salts into the shrinking sea, but with nowhere for the salt to go the water got saltier and saltier. Because circulation with other oceans was cut off, salt levels fell globally by about two parts per million. That seemingly tiny amount was sufficient to raise the freezing point of sea water, which may in turn have been a key factor that caused ice sheets to engulf Antarctica at this time – which meant lower sea levels, prolonging the Mediterranean's isolation.

Evidence of the crisis first came to light in 1961, when a seismic survey revealed that something under the sea floor was reflecting seismic waves in an unusual way. Subsequent drilling revealed a deep layer of salt that had been deposited by the drying sea. This layer is the source of the rock salt mined throughout the region, and it stretches right across the Mediterranean, in places measuring 2 km (1.2 miles) deep. In Sicily, for example, the landscape is littered with mounds of salt pushed to the surface by geological turmoil, and the salt mines in western Sicily are said to contain enough salt to stay in operation for a million years.

It is unclear what brought the crisis to an end. The favourite theory is that rivers eroded the land barrier at the Strait of Gibraltar at the same time as climate change was raising sea levels beyond. And so the ocean flooded back in.

BELOW **IAIN STEWART STANDING AMIDST THE CRUMPLED LAYERS OF SALT LAID DOWN IN THE MESSINIAN CRISIS. SOME MINES IN SICILY HAVE ENOUGH SALT FOR A MILLION YEARS' SUPPLY.**
OPPOSITE **FIVE MILLION YEARS AGO, THE FLOOR OF THE MEDITERRANEAN SEA WOULD HAVE LOOKED LIKE THESE ARGENTINIAN SALT FLATS IN THE ANDEAN ALTIPLANO.**

become snagged together as one tried to shove its way under the other, causing an immense build-up of stress. When the inevitable rupture happened, the locked surfaces jerked violently apart and scraped across each other along the whole length of a 1600 km (1000 mile) fault line, sliding up to 18 m (59 feet) in a few minutes. Much of the movement happened in the overlaying plate as the release of tension allowed its snagged edge to spring back up, causing the sea floor to lurch up by several metres across a vast area. The quantity of energy released was so great that the numbers are meaninglessly huge, but it was sufficient to make the planet wobble on its axis by up to 2.5 cm (1 inch) and accelerate its rotation, shortening the day by 2.68 microseconds. Much of that energy was transferred to the 30 cubic km (7 cubic miles) of water that was displaced, generating a succession of tsunami waves that not only pounded the coasts of Indonesia, Thailand and Sri Lanka but also spread into the Pacific and crossed the Indian Ocean to strike Africa and sweep around the Cape of Good Hope into the Atlantic. Some 230,000 people were killed in the areas that felt the full force of the tsunami, but its reach was global.

By chance, the event occurred precisely as two radar satellites were passing over the Indian Ocean, and the data they gathered made it possible to build an accurate picture of the size and spread of the waves. Tsunamis behave differently in deep and shallow water. Out in the ocean, where the water is deepest, the wave was barely detectable as a gentle hump that sped across the surface at up to 1000 km/h (620 mph) – just a ripple that might have passed unnoticed under ships. But that ripple represented a displacement of the entire water column, all the way to the sea bed, which meant that it carried phenomenal energy. Close to the shore, as the sea became shallower, all that energy was compressed, and as the wave was slowed by friction with the rising sea floor, so

RIGHT **THE TSUNAMI OF BOXING DAY 2004 KILLED 230,000 PEOPLE AND DESTROYED MILLIONS OF LIVELIHOODS.**

enormous forces built up behind it, lifting the water up by 30 metres (100 feet) and driving it inland with such tragic consequences. But apart from the final terrible inrush on to land, there was little forward movement of water in the tsunami – as the wave traversed the ocean, it wasn't water that was travelling but pure energy itself. Indeed this is true of all ocean waves, including the vast breakers that are conjured up by storm winds in the Antarctic Ocean and then travel north across the Pacific for as much as a week before curling on to the beaches of Hawaii or the other Polynesian islands. For what the ocean does so effectively is to transfer energy from one place to another and move it around the globe. It does so in many different ways, all of them crucial to our planet's equilibrium.

Twice every month – at new moon and full moon – Earth, the Sun and the Moon lie in a straight line. When this happens, the gravity of the Sun and Moon act together to pull on Earth's oceans and raise tides to their maximum height, creating 'spring' tides. Every other fortnight, the Sun and Moon lie at right angles, so that the pull of the Moon is partly cancelled out by that of the Sun; result: 'neap' tides. If the world were covered with a uniformly deep ocean, then the Moon would pull the water towards it by an unvarying 54 cm (21.2 inches). In reality, of course, the tide is influenced by a multitude of obstacles from continents and islands to reefs and estuaries, and these obstacles, can sometimes have remarkable effects.

Two neighbouring bays on the east coast of Canada vie for the accolade of having the world's highest tide. Burntcoat Head in the Bay of Fundy has long claimed the crown, with a maximum recorded difference between high and low tides of 17 metres (55.8 feet). But modern instrumentation has recently shown that Leaf Basin in Ungava Bay just to the east has a maximum tidal range of 16.8 metres (55.1 feet). Since the difference between the two measures falls within the bounds of experimental error, the two localities now have to share the title until further measurements can be taken during the next extreme high tide, which will not happen until 2015, when a particular astronomical alignment will exert maximum pull on Earth's oceans. At the opposite end of the scale from eastern

ABOVE **COASTAL EROSION. THE PACIFIC IS GRADUALLY CLAIMING THESE HOUSES IN CALIFORNIA.** OVERLEAF **CROWDS LINE THE BANKS OF THE QIANTANG RIVER IN CHINA TO WATCH THE ANNUAL ARRIVAL OF THE WORLD'S GREATEST TIDAL BORE.**

Canada are the coasts of enclosed seas like the Mediterranean, where there is almost no tide because water cannot get through the Strait of Gibraltar fast enough to make a difference. Meanwhile, in Hangzhou Bay on the east coast of China, incoming tides are funnelled towards the Qiantang River by the bay's tapering shape. A large peninsula of reclaimed land at the mouth of the river constricts the inlet dramatically, and the river bed rises sharply. This unique geographical arrangement regularly creates a breathtaking natural wonder: the world's largest tidal bore. This ferocious wave races into the bay at the speed of a galloping horse and can climb to a height of 9 metres (30 feet) before showering onlookers in a roaring surge. Every September, when the bore is at its highest, a festival is held in the small town of Yanguan on the north side of the river, and thousands throng to the bank to watch the spectacle. But the wave can be dangerously

unpredictable, and many people who stood too close to the edge have been swept to their death.

The vast majority of the waves we see breaking on a beach are caused by wind dragging across the surface of the ocean far way. The energy can travel thousands of kilometres before being unleashed as the wave collides into land and breaks. An average storm wave is said to crash into the shore with the same force as two elephants jumping on your chest at the same time. Unsurprisingly, this amount of energy, pounding shores day in and day out, does great damage to the land and continually reshapes coastlines. Places such as Norfolk in England, where the coast is retreating faster than anywhere else in Europe, are locked in constant battle with the waves. At the village of Happisburgh in north Norfolk, the sea is very clearly winning that battle. The Norfolk coast is made of sand and clay laid down by glaciers at the height of the ice age, and these soft glacial deposits erode readily, allowing the sea to wash away land at a rate of 2 metres (6.5 feet) a year, or even by 10 metres (33 feet) a year in the most exposed areas. About 500 years ago, the coast was a kilometre further out and supported villages that have long since vanished into the water. Today, an average of one house a year is lost to the sea. The old sea defences have at best only slowed the rate of destruction, and they may even have made things worse by disrupting the manner in which sediments are washed along the coast. In the end, the Norfolk coastline will be reshaped dramatically, and the process will accelerate with the sea level rise that global warming will bring.

Coasts around the globe bear spectacular witness to the power of waves. Sitting proudly in the surf at Cannon Beach in Oregon, USA, is Haystack Rock – the world's third largest sea stack. A scattering of smaller rocky shards sits beside it. At one time, Haystack and its

RIGHT **THIS DRAMATIC LIMESTONE CLIFF ON THE COAST OF VICTORIA, AUSTRALIA, HAS ERODED TO LEAVE ONLY EIGHT OF THE FAMOUS 'TWELVE APOSTLES'.**

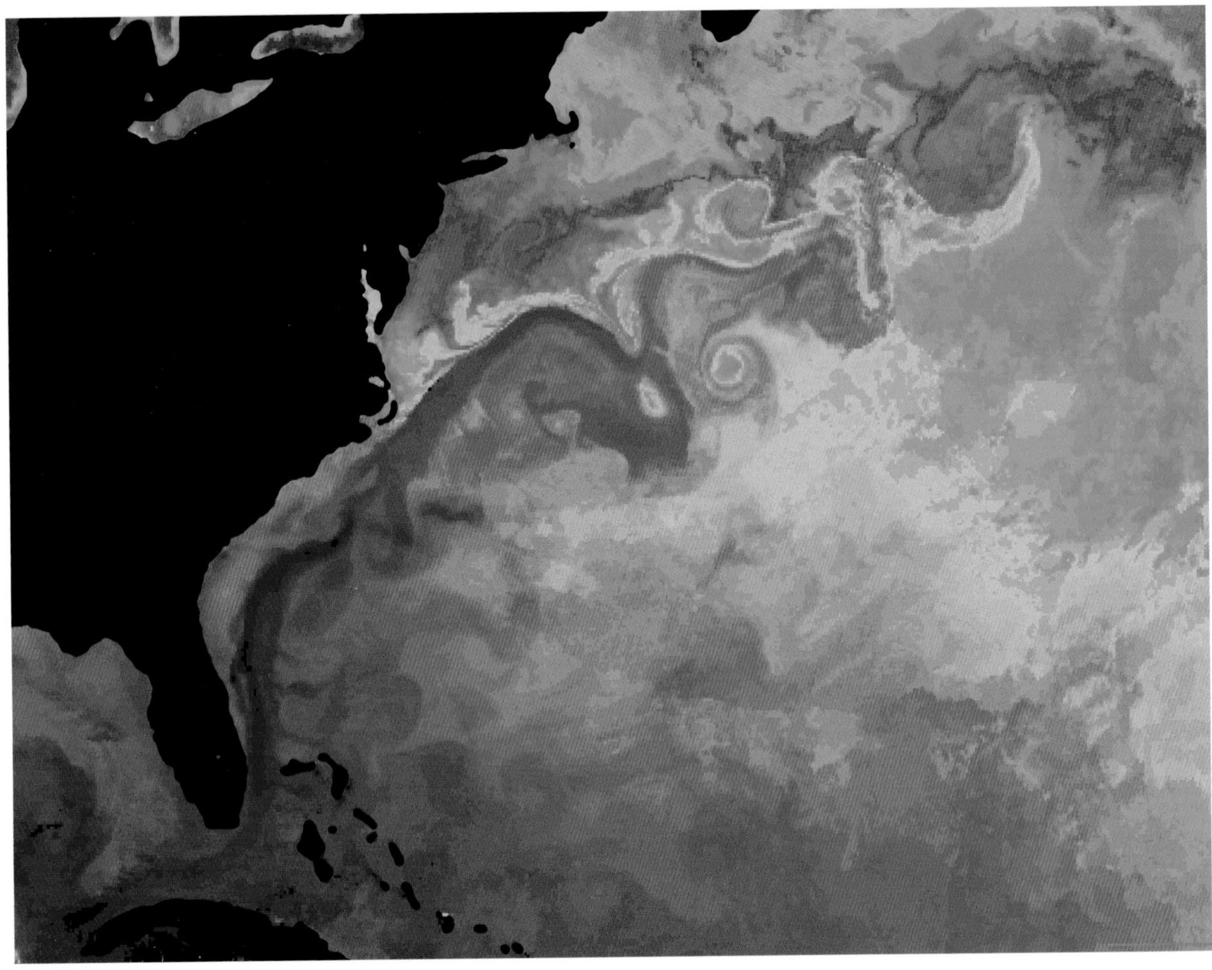

ABOVE **THE TEMPERATURE DIFFERENCE BETWEEN THE WATERS OF THE GULF STREAM AND THOSE ON EITHER SIDE CAN BE AS MUCH AS 10°C (18°F).**

companions were part of the coastal headland, but thousands of years of erosion have destroyed so much land that the rocks now stand isolated more than 100 metres (330 feet) from the shore. But the coast here wasn't made of sand and clay like that at Happisburgh in Norfolk; it was made of basalt – solid volcanic rock – and the ocean has eaten it away as if it were butter. On the coast of Victoria in southeast Australia is an even more impressive demonstration of the ocean's destructive might. The Twelve Apostles are towering needles of rock that were once part of the limestone cliffs running along the coast. After gouging them out of the rock, the sea began eating them away one by one. In October 2005, a pillar of rock 50 metres (160 feet) tall finally crumbled and fell into the sea, its base undermined by erosion. It was the fourth to collapse since the rocks were named. They might still be called the Twelve Apostles, but due to the ravages of the ocean there are now only eight.

Ocean River

Although waves carry only energy and not water itself, there are ways in which ocean waters can move around the planet, in massive quantities, and these also transfer energy through the global system. They are the ocean currents, and perhaps the most famous of them is the Gulf Stream. Emerging from the Gulf of Mexico, this fast-moving warm current sweeps around Florida and flows up the USA's east coast. Its first historical mention was in 1513, when the Spanish explorer Juan Ponce de León (who had been seeking the mythical 'fountain of youth' but had to settle for the discovery of Florida) remarked on its curious effects in his ship's log. Ponce de León was sailing south through the Florida Straits when, despite having a strong tailwind, his vessel began to move backwards. He put it down to a current caused by the narrow strait, and he steered for the shore to get out of its path. What he did not realize was that this gigantic river of sea water, which flows through the Florida Straits at a rate of 30 million tonnes a second, could have carried him all the way back to Europe at speeds of up to 160 km (100 miles) a day.

Spanish sailors were soon taking advantage of the short cut across the Atlantic, but they kept its existence to themselves. The atmosphere of secrecy continued until the late eighteenth century, when the Gulf Stream was finally charted by none other than Benjamin Franklin, acclaimed scientist, diplomat and Founding Father of the United States. Franklin is perhaps best known for flying a kite into a thundercloud to test his electricity theory, but what is less well known is that he spent part of his life in England working as deputy postmaster general for the American colonies. He was perplexed as to why mail packets should take two weeks longer than merchant ships to sail to America, so he asked his cousin Timothy Folger, a whaling captain, for advice. As Folger explained, those in the know were steering a course to the south to avoid the Gulf Stream, but the mail boats were unwittingly sailing against the current. Folger knew all about the secret current because whales would often gather to feed along its edges, and he had told mail ship captains to alter course; but the words of a mere whaler fell on deaf ears. He now drew a chart of the 'stream' for Franklin, who had it printed for the postmaster general – who of course ignored it. But Franklin was determined to find out more, and in 1775 he sailed across the Atlantic taking temperature readings from the side of his ship. He discovered that the current was unusually warm, with blue water unlike the normal Atlantic grey. Whales did not swim into it because it appeared to carry little food. Franklin even found time to continue these scientific observations on every day of the perilous voyage he made to France in 1776 to rally support for the American Revolution. His studies were fruitful, and he pretty much correctly explained the current as being due to an 'accumulation of water on the eastern coast of America between the tropics', driven there by the trade winds. He also suggested the name Gulf Stream.

Today it is one of the best-studied ocean currents in the world. It has been probed, tracked, measured, imaged and analysed in every conceivable way by a mind-boggling array of instruments: neutrally buoyant current meters, acoustic SOFAR and RAFOS floats, echo sounders, tomographic transceivers, infrared satellites, microwave satellites, and many more. The picture that has emerged is wondrous in its complexity. The Gulf Stream is more than just a warm river – it is a whole circulatory system, spawning vast eddies of water that spin off and loop back again. Earth's rotation bends the whole stream to the right (the Coriolis effect), but because the southeast side is a metre (3 feet) higher than the northwest, gravity tries to heave the vast mass of water back the other way. The current hugs the USA's southeast coast, measuring only 80–150 km (50–90 miles) wide, its flow rate rising as it travels north and reaching a peak of 150 million cubic metres (5.3 billion cubic feet) a second, or 540 billion tonnes of water per hour – whichever statistic is easier to comprehend in its vastness. From the northeast USA it heads out across the Atlantic and forks. The northern fork continues towards Scotland and Norway as the North Atlantic Drift, and the southern fork turns clockwise to head back towards the Caribbean.

Calling Rubber Duck

On 10 January 1992, an ocean freighter travelling from Hong Kong to Tacoma, USA, ran into severe weather in the North Pacific. The ship was rolling from side to side, tilting up to 40 degrees, and although it came through unscathed 12 of its containers were washed overboard. Inside one of them was a consignment of 29,000 plastic bath toys in little packets, each containing a yellow duck, a red beaver, a blue turtle and a green frog. The packaging fell apart within hours of entering the sea, turning the little creatures loose to set off on an epic journey. Ten months later, when ducks began beaching near the town of Sitka in Alaska, oceanographers realized they had a remarkable opportunity on their hands.

The study of ocean currents is not easy, for the precise movement of a given patch of sea water is nigh on impossible to follow. Electronic floats that can be tracked by satellite cost around £2500 each, and they have a short life in the ocean. But flotsam and jetsam are free. Some 10,000 cargo containers go overboard around the world every year, and scientists have in the past tracked the voyages of Lego building blocks, hockey gloves, umbrella handles and, a particular favourite that turns up from time to time, Nike trainers.

The plastic bath toys' initial travels took them to Alaska, west to Japan, and back to North America over a period of three years. Then they ran into the North Pacific Gyre – a swirling vortex of currents that encircles most of the North Pacific. A few were caught in the centre of the gyre in an area known as the 'great Pacific garbage patch', where floating debris accumulates in slow waters, and those unlucky ducks and their friends will end their days there, gradually degrading in this disgusting and most polluted part of the world's oceans. However, others were spun away in different directions. Thousands floated through the tropics, following ocean currents that took them across the equator to Indonesia, Australia, and back over the Pacific to South America. About 10,000 went north through the Bering Strait between Alaska and Russia and into the Arctic Ocean. Some of these became frozen into sea ice and were carried to the pole before turning south to pass Greenland and Iceland, finally being released into the Atlantic by melting ice floes. By 2003 a green frog had been found in the Hebrides, while others had reached the east coast of North America. Still others travelled south to the West Indies, where they were caught in the fast-moving Gulf Stream and brought back east to mainland Europe and the UK.

If you find one, look for the words 'The First Years' embossed on the side – you might be able to claim a $100 reward.

BELOW **A SIMPLIFIED ROUTE OF THE GLOBAL CONVEYOR'S THOUSAND-YEAR CYCLE THAT LINKS ALL THE OCEANS ON THE PLANET.**

The Gulf Stream carries a phenomenal quantity of heat energy, calculated to be 100 times the entire energy consumption of the human race, or roughly equivalent to the output of a million power stations – either way, a humbling thought. It is certainly the kind of energy that can influence the climate of northern Europe, giving us much milder winters than we'd otherwise have and generating the weather fronts that keep us well supplied with cloud and rain.

It is not only on Europe's weather that the Gulf Stream has a dramatic influence. To the south of the current is the Sargasso Sea, a huge volume of warm water trapped within a circulating eddy that spins out from the main current. Warm air builds up over the Sargasso Sea, and the Gulf Stream creates a sharp boundary between this warm air and the colder air over continental North America. This boundary is the site of some extraordinary weather phenomena. In winter, cold Arctic air slips gently down across North America before spilling out to sea and encountering the warm, humid air over the Gulf Stream. This results in huge chimneys of steam rising from the surface of the sea to reach the bottom of the clouds. The energy bound up in these 'steam devils' is vast – equivalent to a nuclear power station pumping energy into the atmosphere from every square kilometre of the sea. The astonishing updraughts can have far-reaching effects, from fuelling ice storms over New England to building the Atlantic cyclones that drench the British Isles.

Along the so-called north wall of the Gulf Stream, warm, moist air rises and flows out across water that is much cooler, creating large banks of fog. For half the year, fog also forms where the icy Labrador Current swings south to strike humid air that has crossed the Gulf Stream, the sudden chill causing the moisture to condense out. Icebergs drifting south from Greenland add to the fog, only to emerge from it suddenly when they reach the clear air of the Gulf Stream, where they are caught and melt before they can go further south.

Global Conveyor

Surface currents such as the Gulf Stream are driven by the wind and can move with surprising speed. But in the last 30 years or so, scientists have also discovered much slower streams of water winding across the ocean floor. And all these currents turn out to be interconnected, forming a global circulatory system that carries not just water around the globe but vast amounts of heat too, with profound consequences for the planet's climate.

Imagine a time, long ago, when a sailing ship was nearing its port in western Europe, blown along by the Atlantic westerlies after a journey of courage, adventure and privation that few of us can comprehend today. Imagine, perhaps, a sailor on one of Sir Francis Drake's privateers, possibly the great sea dog himself, sluicing his face to freshen up in the early morning and musing on the clean clothes his newly plundered wealth might afford him when he reaches Plymouth in a few days' time. Imagine the dirty water tipped unceremoniously over the side of the ship to mix and swirl away in the foaming brine of the ocean. Well, that very water may rain upon you tomorrow, for Sir Francis Drake's ablutions would certainly have become a tiny part of the greatest ocean current of all: the thermohaline circulation, known more commonly as the global conveyor.

It is a continuous loop, but let's choose a starting point in the North Atlantic Ocean, where Sir Francis threw his dirty water overboard. First the water heads north, carried by the Gulf Stream's northern fork: the North Atlantic Drift. This current is warm and salty (in fact the North Atlantic is the warmest and saltiest of the oceans), but near the Arctic it cools, becoming heavier, and sea ice forms on its surface. The ice draws only on pure water to build its frozen structure, leaving salt behind. Now the remaining water is saltier as well as colder, making it even heavier, and so it sinks. Only visible through sonic imagery, this mass of cold, dense, salty water plummets to an extraordinary depth, dropping three and a half times further than Angel Falls, the tallest waterfall on land, to join what oceanographers have catchily named the North Atlantic Deep Water – a mass of dense, saline water that pools in the Atlantic basin. From here, Sir Francis's water begins a long, sluggish journey south, through the cold darkness of the ocean depths, until it reaches the other side of the world.

The long journey is not straightforward. The water cannot cross the ocean directly – it must travel in stages, much as water from a fountain passes down a cascade, swirling around in each basin for a while before spilling to the next level. The deep waters of the global conveyor move through a series of 'gyres' – rotating currents that can be as wide as the whole ocean, with water cycling slowly around each one for years before slipping out to the next one. It might take Sir Francis's water a century to circle round to South America, perhaps reaching northern Brazil shortly after the Great Fire of London. From there it continues its circuitous journey south towards Antarctica, arriving another century later, perhaps at the height of the French Revolution's reign of terror.

In Antarctica the water is joined by a cold current descending off the Antarctic coast, and then the conveyor splits in two, one branch circling Antarctica for Australia and eventually, a millennium later, resurfacing in the North Pacific. The other branch heads north – travelling up East Africa as the era of Victorian exploration is getting under way – and gradually warms, decade by decade, until it surfaces near India at the end of the Raj. Now at the surface, the journey speeds up. With the wind behind it, the water heads south again for Africa's Cape of Good Hope, swings around it, and then crosses the Atlantic for the Caribbean – though on the way it spends a few years stuck here and there in spiralling eddies and gyres. Once again it enters the Gulf Stream, to be swept with remarkable speed back to the North Atlantic, from where it evaporates, drifts

RIGHT **THE OCEAN VORTICES GENERATED BY THE ANTARCTIC WINDS IN THE WEDDELL SEA ARE AN ESSENTIAL PART OF THE CIRCULATION OF THE GLOBAL CONVEYOR.**

Boomerang Man

In 1899, India was crippled by famine after the monsoon rains failed. The Indian Meteorological Service appointed a new director general, and he came to office determined to understand the cause and figure out a way of predicting the monsoon. His name was Sir Gilbert Walker and he was a classic Edwardian polymath with an extraordinary range of artistic and scientific interests, including painting, ice-skating, gliding and the scientific study of bird flight. Walker was fascinated by the performance of the flute and made a small improvement to its design, and he spent 10 years working on the properties of the boomerang (countless examples of which he'd had shipped over from Australia) and using it to develop new theories about the nature of gyroscopic motion.

Walker's approach to the monsoon problem was to apply the art of statistics. This was long before the days of number-crunching computers, but Walker had plenty of manpower at his disposal. He set his many highly trained assistants in the Indian Meteorological Service the task of scouring weather records from all over the world for any kind of statistically significant correlation between major meteorological events. And a link emerged: Walker found that the monsoon's severity and timing were related to the relative air pressures over the Indian and Pacific Oceans. He concluded that the monsoon must be just one part of a greater global weather system, and he named the oscillating air pressures over two oceans the southern oscillation.

Walker spent years amassing and analysing data and found that the oscillation was also linked to rain and wind patterns in the Pacific and Indian Oceans and to temperature changes in Africa, southern Canada and the USA. But his attempts to predict the monsoon still failed, and his evidence of a link was called into question by other meteorologists. Years later, it is now obvious that Walker was very much on the right track, and the pattern of circulating air over the equatorial Pacific was named the Walker Circulation in his honour – for what Walker achieved was far more than merely predicting the monsoon: he laid the foundations for studying global climate as a single, interconnected system.

ABOVE **THE EDWARDIAN METEOROLOGIST SIR GILBERT WALKER.**
BELOW **THE REGULAR MONSOON RAINS OF ASIA ARE JUST ANOTHER PART OF LIFE.**

over Europe, and rains back down in a spring shower.

This extraordinary and epic journey of water is vital to the workings of our planet in many different ways. Working in partnership with the atmosphere, the global conveyor redistributes heat around the planet, evening out the extremes of sweltering tropics and freezing poles. The sinking cold water carries great quantities of fresh oxygen to the depths – for cold water holds more oxygen than warm – and turns what would be a stagnant abyss into a haven for life. On its slow meander across the sea floor, the water picks up nutrients from sediment and from the 'snow' of dead plant and animal matter falling from sunlit waters far above. In the Southern Ocean, the howling winds that encircle Antarctica shift the top waters, allowing some of the deep water to rise. Antarctica's land may be a barren wilderness, but the nutrient-rich upwellings make its seas some of the most fertile on the planet. Huge blooms of plankton thrive there, providing sustenance for an entire marine ecosystem, the nutrients from the deep sea passing up the food chain to krill, anchovies, penguins, seals and whales.

Driven by the simple physics of rising warm water and sinking cold, salty water, the global conveyor keeps the health of our planet in check, and it mostly takes place without anybody noticing. That is, until it goes wrong.

A Cooler, Warmer World

In the 1960s, British geologist Russell Coope began to excavate dead ground beetles from layers of ancient mud in a cliff at St Bees on the northwest coast of England. Thousands of years ago, the carnivorous beetles had lived happily around the edge of a muddy pool, failed to look where they were going, and fallen in to meet their doom. Year after year the beetles tumbled in, the mud settling on top of them to form layer after layer of sediment, preserving their bodies for posterity.

Ground beetles are particularly sensitive to temperature, and each species is restricted to a certain climatic zone. But the species in the mud at St Bees varied with time, so the climate must have been changing. Coope excavated as many as he could and found that, by identifying them, he could build up a remarkable record of past climate stretching back thousands of years. But something didn't make sense. Around 11,500 years ago, when the world was emerging from the ice age, warm-loving beetles should have gradually replaced the cold-loving ones. To Coope's surprise, the opposite happened. In fact, the climate appeared to swing back and forth dramatically: warming up, then suddenly freezing, then warming again. Some 11,500 years ago, the warming world must have plunged back into a mini ice age.

Coope's work remained unnoticed or disbelieved for almost 30 years. The standard view was that the ice age ended gradually, and a few dead beetles weren't going to change that. But then came the first results from tests on ice cores drilled from the ice sheets of Greenland and Antarctica. Ice cores can reveal a great deal about past climate. Bubbles of air trapped in ancient layers of snow provide tiny samples of the prehistoric atmosphere, and isotopes in the water reveal how temperatures have changed. The picture that emerged was not one of gradual warming but of sudden shifts in terrifyingly short periods, with global temperatures changing by up to 10°C (18°F) in as little as a decade. And what Coope had seen was mirrored in the ice. Around 11,500 years ago, when the world was generally warming, there was a sudden return of ice age conditions. The cold snap, now called the Younger Dryas (after a tundra plant that flourished across Europe at the time), lasted more than 1000 years. But how could the climate change so rapidly? And could it happen again?

One theory is that the global conveyor broke down. A sudden rush of meltwater from collapsing ice-dammed lakes in North America could have flowed into the path of the Gulf Stream, diluting the salty water and stopping it from sinking. And if the conveyor stopped, the Gulf Stream would have weakened and

shifted south, robbing the Atlantic region of warmth. Whether the Younger Dryas happened this way is hotly debated, but climatologists are convinced the conveyor has switched off many times in the past, and they think that global warming could switch it off again.

In a warmer world, rainfall will increase in the North Atlantic, diluting the sea with fresh water. Melting glaciers will add more. No-one knows how quickly things might happen, but the conveyor is thought to be sensitive to change, and the possibility of it shutting down is no longer remote. If it did, Britain's average winter temperature would fall 2–5°C (3.6–9°F), making the average winter as severe as the coldest year since records began almost 250 years ago. Ordinary winters would be colder than the 'great winter' of 1962–63 (the coldest of the last century), and much colder winters could be expected – many of them worse than the famous winter of 1683, when the Thames froze over with ice 30 cm (1 foot) thick and sea ice blocked the English Channel. Strange as it may seem, global warming could usher in a new ice age.

The Christ Child

The intricate relationship between ocean, atmosphere and climate is perhaps best seen in the infamous meteorological event known as El Niño. In the last two decades, El Niño has become one of the most studied climate phenomena on the planet, because it is also one of the most feared.

Shortly after Christmas each year, a warm ocean current flows south along the coasts of Ecuador and Peru. Occasionally, that current is stronger, flows further south and is very warm, bringing extra-heavy rain which the coastal population has traditionally welcomed for the abundance of crops that follows. Hence the name El Niño (Spanish for 'little boy' in reference to the Christ Child) for this unusual warm current,

LEFT **EL NIÑO BRINGS PLENTY TO SOME PARTS OF THE GLOBE, BUT DEVASTATION AND INUNDATION TO MANY OTHERS.**

which heralds a season of plenty so soon after the Nativity. But El Niño is not always a cause for celebration, for it can also bring destructive weather to many parts of the world, causing floods, drought, a reversal of normal rainfall patterns, and widespread misery and ruin. In 1998, one of the strongest El Niños on record brought chaos to Peru, where 30,000 homes were destroyed by floods and a 150 km (93 mile) long lake appeared almost overnight in a desert that had been dry for more than 15 years.

To understand a major El Niño event, we need first to understand normal conditions. Just off the Pacific coast of South America there is usually a high pressure area from which trade winds blow west across the Pacific, towards low pressure over Indonesia. These steady winds are strong and push the warm equatorial surface waters westwards, piling them in Southeast Asia, where the sea level is some 60 cm (24 inches) higher. The warm current makes Southeast Asia's air more humid, adding to the monsoon rains. In contrast, coastal South America has an arid climate. Because surface waters are pushed away from the coast by the wind, cold, nutrient-rich water wells up from the deep, making the sea here much cooler and the air less humid. While surface winds blow west towards Asia, higher winds flow back the other way, recirculating air after it has shed its rain and so bringing cold, dry air back to the South American coast to complete the circuit; this atmospheric cycle is known as the Walker Circulation (*see* 'Boomerang Man', page 162).

It's a very finely balanced system, and it takes only a tiny change in conditions for the whole thing to break down, leading to an El Niño event. The trigger is a slight relaxation in the trade winds, which immediately causes a feedback loop to set in. The mass of warm water piled high in the west suddenly surges back east, taking warm water and humid air closer to South

RIGHT **EL ÑINO'S DISRUPTION TO THE CLIMATE SYSTEM RESULTS IN EXTREMES OF WILDFIRE AS WELL AS FLOOD.**

America than usual. That, in turn, warms the air and so lowers the air pressure, weakening the winds even more. Suddenly there is a reversal of the Pacific's usual weather, with coastal South America becoming rainy while Southeast Asia and Australia are hit by drought. For farmers in Peru the rains are a boon, but for the fishermen who normally haul in the world's greatest anchovy catches, El Niño is a disaster. Without strong winds to generate the upwelling of water along the coast, the sea loses its main source of nutrients and the fish flee south or die.

The effects of a major El Niño are widespread. On the west side of the equatorial Pacific, the normally humid jungles of Southeast Asia become tinder dry, and monsoons may fail to water the land as far away as western India and the Ethiopian highlands. The tropical storms that normally drench Southeast Asia move east to soak arid Pacific islands. The tops of the vast storm clouds reach high into the path of the jet stream (the fast-moving, high-altitude wind that encircles the planet) and deflect it from its normal path, with consequences for weather around the globe.

To the west, El Niño brings incendiary droughts, wildfires and famine; to the east it dispatches catastrophic floods. These massive climatic upheavals seem to occur every 3–7 years and last for around 18 months. Then, as El Niño dies, the pattern often swings to the opposite extreme, with unusually cool conditions in the eastern Pacific and devastating floods in the west: a mirror image of El Niño known as La Niña ('the little girl'). This huge ebb and flow of warm water across the Pacific is almost unimaginably vast – think of it as a wave the size of the USA sloshing slowly from one side of the planet to the other, perfectly fitting its full scientific name: the El Niño southern oscillation (ENSO).

For Earth, El Niño is merely a hiccup – a momentary redistribution of heat and energy that kicks in to rebalance the global climate system. On a human scale, however, it can be the fifth horseman of Apocalyptic disaster, bringing devastating bouts of climate chaos. In recent years, historians have begun to see the hand of El Niño in great, sometimes epochal, historical events, with grand claims being made for its impact on civilization. The French Revolution, the Irish potato famine, and the agricultural implosion of Chairman Mao's Great Leap Forward all coincided with ruinous harvests and unusually wet, cool summers – conditions that climatologists today attribute to El Niño. And there are signs in the geological record of a spate of super El Niños, some 20 per cent more powerful than the most severe of recent years. These seem to have happened every 300–500 years in medieval times, and they are associated with cataclysmic droughts and fires in the Amazon and floods of biblical proportions in coastal Peru. There are signs that such super El Niños will happen again, because global warming seems to be speeding up and intensifying the ENSO seesaw. In almost three centuries of historical records of ENSO events, there are only eight or nine 'very strong' El Niños, an average of one every 42 years, yet two of the largest occurred in 1982–83 and 1997–98, an interval of only 14 years.

When Oceans Go Bad

In the previous chapter we saw how life on Earth came close to total annihilation some 250 million years ago, at the end of the Permian Period. Evidence of this mass extinction can still be found today – in some surprising places. At an altitude of 2000 metres (6500 feet) in the Italian Alps is a layer of black rock that was once part of the ocean floor. Tectonic movement has raised this ancient sea bed to become part of Italy's snowcapped Dolomites, a spectacular range of craggy limestone peaks that stand over picturesque valleys cut by waterfalls and streams. The black rock is called shale, and compressed within it are the victims of the great wipe-out, when 96 per cent of all marine species perished.

The volcanic eruptions in Siberia that seem to have triggered the extinction tore open gigantic rifts in Earth's crust, forcing out giant curtains of bubbling lava and spewing carbon dioxide into the atmosphere almost continually for a million years. The resulting greenhouse effect raised global temperatures by an estimated 5°C (9°F). But it was in the ocean that the full consequences played out. Just as today, the warming was greatest at the poles, and as polar waters became too warm to sink the global conveyor would have shut down. Deep ocean currents ground to a halt, depriving the deep sea of oxygen, and the surface water became warm all over the world. Below a few hundred metres, the water became stagnant, warm and motionless; below 1000 metres (3300 feet), everything died.

ABOVE **THE FOSSILIZED JAW OF A RHINESUCHUS, AN AMPHIBIAN THAT WAS WIPED OUT, ALONG WITH ALMOST ALL LAND AND MARINE SPECIES ON EARTH, AT THE TIME OF THE PERMIAN MASS EXTINCTION.**

The knock-on effect of so many deaths in the deep sea was a collapse of the entire marine food chain, and in time most ocean life was killed off – through lack of oxygen, lack of food or both. The remains slowly sank to the bottom of the sea, piling layer upon layer, to rot. Except that they could not even do that, for the oxygen-dependent bacteria that normally strip carcasses on the ocean floor were denied oxygen too. So the remains just stayed where they were and slowly compressed into a thick, black, slimy sludge, eventually to become shale. It was a global disaster. The 250-million-year-old shale has been found on every continent – in China, America, South Africa and the Arctic – and wherever you find it, you know you're looking at the remains of billions of marine organisms killed by lack of oxygen.

Land animals fared little better, and you can see why by visiting a remarkable lake in upstate New York, USA. Surrounded by thick forest, Green Lake is a popular beauty spot known for the vivid blue-green of its water. But appearances can be deceptive, for exploring this lake is like diving back in time by 250 million years to enter the dying seas of the Permian. The lake is very deep – more than 60 metres (200 feet) in places – and native Americans once believed it to be bottomless. It is also largely cut off from inflowing water. These factors conspire to stop the water circulating, with the result that its depths are devoid of oxygen and as stagnant as a Permian ocean.

If you dived at Green Lake – which would be very unwise without a great deal of protection – you'd see the water turn from clear to a disturbing shade of pink some 20 metres (65 feet) down, as you crossed the boundary at which it becomes stagnant. There may be no oxygen down here, but there are bacteria that can 'breathe' other substances dissolved in the water, the most plentiful of which is sulphur. Just as we breathe oxygen, these bacteria absorb sulphur and use it to obtain energy, and in the depths of Green Lake they are the king of the heap. Switch on a torch and vast numbers of sulphur-loving bacteria glow purple, appearing to dye the water. This is what the Permian deep ocean would have looked like, and it is a telltale sign of the presence of hydrogen sulphide gas, one of the most toxic substances we know of. At high concentration, one lungful will kill you more surely than cyanide, and it is a horrible death. It attacks multiple organ systems at once, but it is especially damaging to the nervous system, paralysing and asphyxiating its victims.

Despite their minuscule size, these bacteria once ruled the seas. In the stagnant Permian oceans, they created vast quantities of hydrogen sulphide – so much that the water became saturated and began to belch it into the air in toxic plumes, wafting the stink of rotten eggs far and wide on the wind. The toxic seas covered 75 per cent of the planet, so the quantity of hydrogen sulphide produced must have been phenomenal. And as hydrogen sulphide flooded the atmosphere, every animal that tried to breathe the poisoned air would have been killed, from the smallest insects to the largest reptiles that lived on land. If that were not sufficient, the hydrogen sulphide also rose into the upper atmosphere and attacked the ozone layer, exposing the plants and animals below to the ravages of high-frequency ultraviolet radiation.

OPPOSITE **CORAL REEFS TEEM WITH MORE FORMS OF LIFE THAN WE HAVE YET DISCOVERED, BUT THEY RISK BEING DESTROYED BY THE OCEAN ACIDIFICATION THAT COMES WITH GLOBAL WARMING.**

If the theory is right, then what happened at the end of the Permian Period was the greatest mass poisoning the world has ever seen. The distant ancestors of those sulphur-loving bacteria in Green Lake were simply unstoppable. They took over the oceans and pumped out lethal chemicals until they had poisoned the entire planet. A few small animals survived – perhaps because they lived deep underground or in a few parts of the world that escaped the poison gas clouds – but life on Earth was devastated, the victim of a chain of events that the oceans had nurtured and amplified.

Ocean Rising

Even though humans evolved on land, the sea continues to dominate our existence, giving us food and oxygen and regulating our weather and climate. The truth is that the world ocean contributes more towards the well-being of our planet than any other part of the global system. So it is all the more worrying that the ocean is showing signs of change.

Sea levels are rising as the oceans warm. This is partly caused by ice melting on land, but it is mostly due to straightforward thermal expansion of water as it gets hotter. For most of the last 3000 years, sea levels have risen about 0.2 mm (0.01 inches) a year, but in the last two centuries the rate has accelerated and they are now rising 2 mm (0.1 inches) a year – ten times as fast. We are now all familiar with the dangers that rising waters pose to coasts, river deltas and low islands around the world.

Another telltale sign of warming is the increase in cases of coral bleaching. Coral organisms owe their vibrant colours to the marine algae that live symbiotically within their soft tissue, providing sugars and proteins in exchange for shelter and protection. But these algae are highly sensitive to temperature, and if it gets too warm they stop producing nutrients. The coral then spits them out, revealing the ghostly white colour of its chalky skeleton, and eventually it dies from lack of

PHYTOPLANKTON

Our distant ancestors may have left the ocean hundreds of millions of years ago, but we are still totally dependent on it – as is every other creature on the planet – because of a group of single-celled organisms known as phytoplankton.

Phytoplankton are probably the most important life forms on Earth, and they dominate the seas, far out-numbering all other forms of marine life. Individually, they are far too small to see with the naked eye, but en masse they are clearly visible from space, flourishing in their trillions to form gigantic, milky blooms that float on the ocean surface. This huge explosion of life is the basis of the marine food chain, supporting nearly all the ocean's inhabitants from the tiniest crustaceans to the largest whales.

But phytoplankton play another role, one that is even more important to the health of our planet: they help the world to breathe. A clue as to how is found in their colour. Blooms of phytoplankton turn the ocean a light emerald colour, for these organisms contain chlorophyll – the same substance that makes plants green and enables them to photosynthesize. Just like plants, phytoplankton soak up the Sun's energy and use it to join molecules of water and carbon dioxide together to synthesize food, releasing oxygen as a by-product. Ironically, the one place on Earth where we can't breathe unaided is the very place that supplies half our oxygen: the ocean. Phytoplankton produce as much of the oxygen in the atmosphere as all the world's forests and jungles combined.

The benefits don't stop there. Phytoplankton pull vast quantities of carbon dioxide out of the water as they photosynthesize, which in turn draws carbon dioxide out of the air, reducing the greenhouse effect. In fact, the oceans have absorbed a third to a half of all the carbon dioxide produced by human activity since the start of the Industrial Revolution. Some species of plankton construct microscopic carbonate skeletons, trapping the carbon, and when they die and sink to the sea floor as 'marine snow', the carbon goes with them. Eventually it turns into limestone and is subducted deep into Earth's crust, from where the carbon can escape through volcanoes, returning to the atmosphere and keeping Earth's global thermostat (see chapter two) in operation.

Phytoplankton need certain minerals, particularly iron, which they get from wind-blown dust falling on the sea. But patterns of land use have reduced the amount of iron carried by the wind, and phytoplankton populations have fallen by at least 6 per cent in the last 25 years. Some scientists propose artificially 'seeding' the oceans with iron particles to boost phytoplankton numbers back up to and beyond their normal level, which would offset greenhouse gas emissions. It may be a far-fetched idea, but in a world that is hotter than it has been for 400 years, the oceans are leading the fight against global warming and they may need all the help they can get.

BELOW **MICROSCOPIC IMAGES OF PHYTOPLANKTON REVEAL THE VARIETY OF SKELETAL STRUCTURES THAT THEY BUILD – ALL CONTRIBUTING TO THE CAPTURE OF CARBON DIOXIDE FROM THE ATMOSPHERE.** OPPOSITE **A 'BLOOM' OF PHYTOPLANKTON OFF THE COAST OF BRITAIN AND FRANCE.**

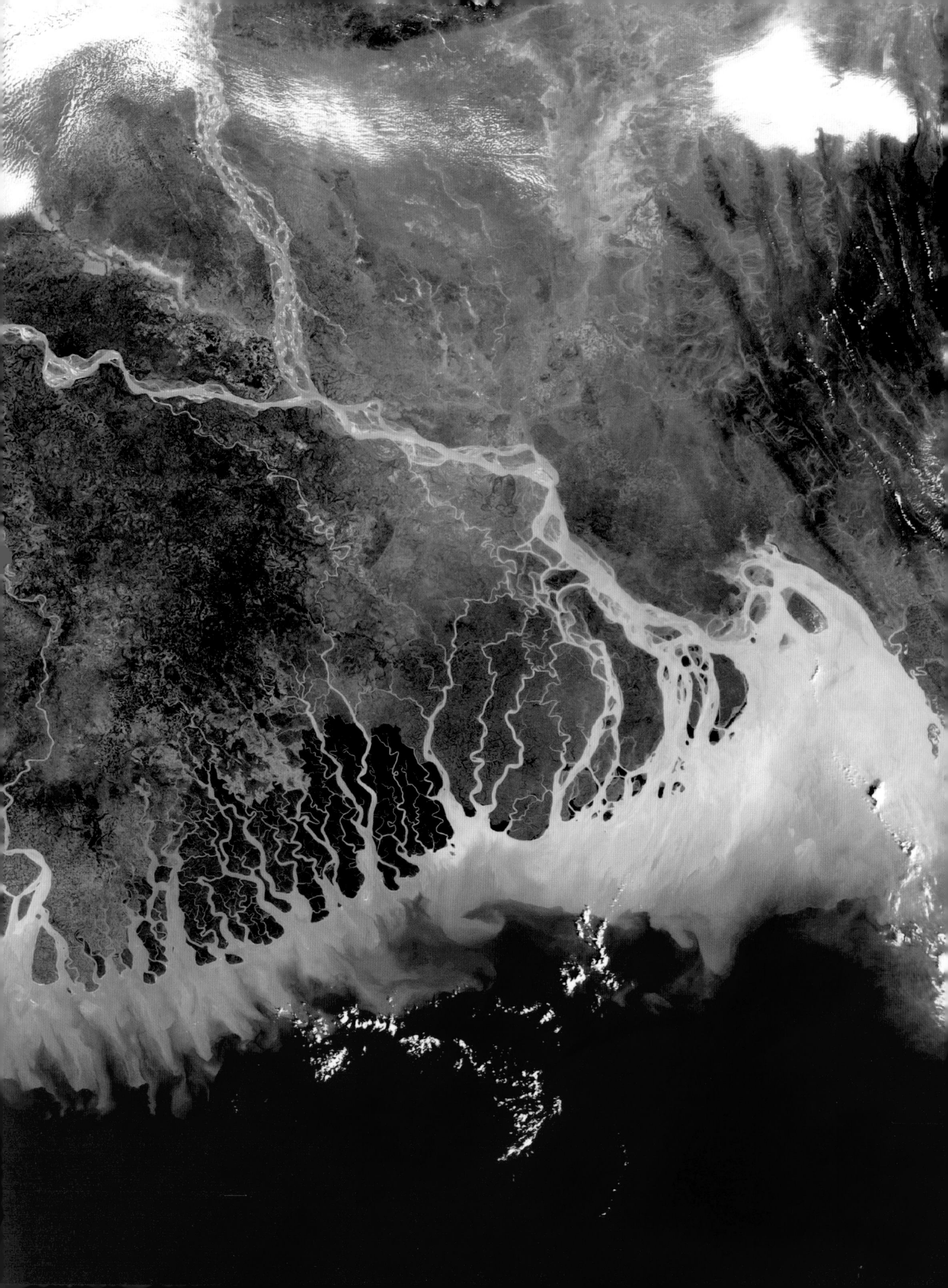

ABOVE **INVESTIGATING THE OCEAN USING A SCUBA-SCOOTER.** OPPOSITE **SATELLITE IMAGE OF THE LOW-LYING DELTA REGION OF BANGLADESH. THESE LANDS WILL SOON BE SUBMERGED AS THE PLANET WARMS AND THE OCEAN EXPANDS.**

nutrients. The Great Barrier Reef off Australia suffered severe bleaching events in 1998 and 2002, with almost half of its coral turning white in the latter episode, from which it has still not recovered.

Even the chemistry of the oceans is changing. As we saw in chapter two, carbon dioxide partially dissolves in water to form a weak acid. So as we pump ever more CO_2 into the atmosphere, the oceans will become slightly more acidic. The consequences of this are not yet fully understood, but one likely outcome is that phytoplankton will find it harder to calcify their fragile skeletons. And because of that, less carbon will be carried to the bottom of the ocean in marine snow, and less carbon dioxide will be absorbed by the sea in the first place, because phytoplankton numbers will be depleted. And just to top things off, when phytoplankton numbers fall, the sea loses the pale colour that reflects the Sun's heat away, so the water will become warmer still – one of the many feedback effects that scientists are trying to come to grips with as they unravel our planet's complex workings.

Perhaps the most worrying part is that we just don't know what will happen. For the past 20 million years or so, the world ocean has enjoyed a period of calm stability and has stayed in balance with the rest of the planet. But today it is being forced to change at a speed that may be unprecedented in geological history, and no-one really knows what the repercussions might be. One thing is certain, however: humans have started something that is irreversible in our lifetimes. Even if we stop polluting the planet right now, restoring the ocean to prehuman conditions could take millions of years.

CHAPTER FIVE

Ice

The moons of Jupiter were discovered by Galileo Galilei in January 1610, and their apparent orbit around the giant planet became a critical point of argument in his battle with the Church over whether Earth or the Sun were at the centre of the solar system. Each of the four Galilean moons has proved to have remarkable properties, and one of them – Europa – offers a vision of what our own world may once have looked like. For Europa is entirely covered in ice. Images from NASA's *Galileo* spacecraft, which orbited Jupiter for six years around the turn of the millennium, have revealed this moon of Jupiter to be a glittering ball of frost, its surface covered entirely by a crust of ice some 30 km (19 miles) thick and patterned by crazed lines where fractures have opened and frozen shut again as the surface moves. It is thought that Europa's ice moves independently of the rocky surface below, rotating once every 10,000 years relative to the solid interior and gliding over a hidden ocean of liquid water as the ice is tugged and stretched by Jupiter's huge gravitational pull. Earth was never as cold as Europa, which sits five times further from the Sun than us and has only the merest trace of an atmosphere to warm its surface. Yet, deep in our planet's history, extraordinary changes in climate have plunged Earth into the freezer too, turning our home into a vast snowball, its icebound surface a dazzling white in the Sun's glare.

OPPOSITE ICE BREAKING OFF THE LEADING EDGES OF THE WORLD'S GLACIERS IS A CONSTANT PROCESS. BUT IT IS SPEEDING UP. OVERLEAF GLOBAL WARMING IS GREATER AT THE POLES THAN IN THE TROPICS. THE POLAR ICECAPS ARE MELTING.

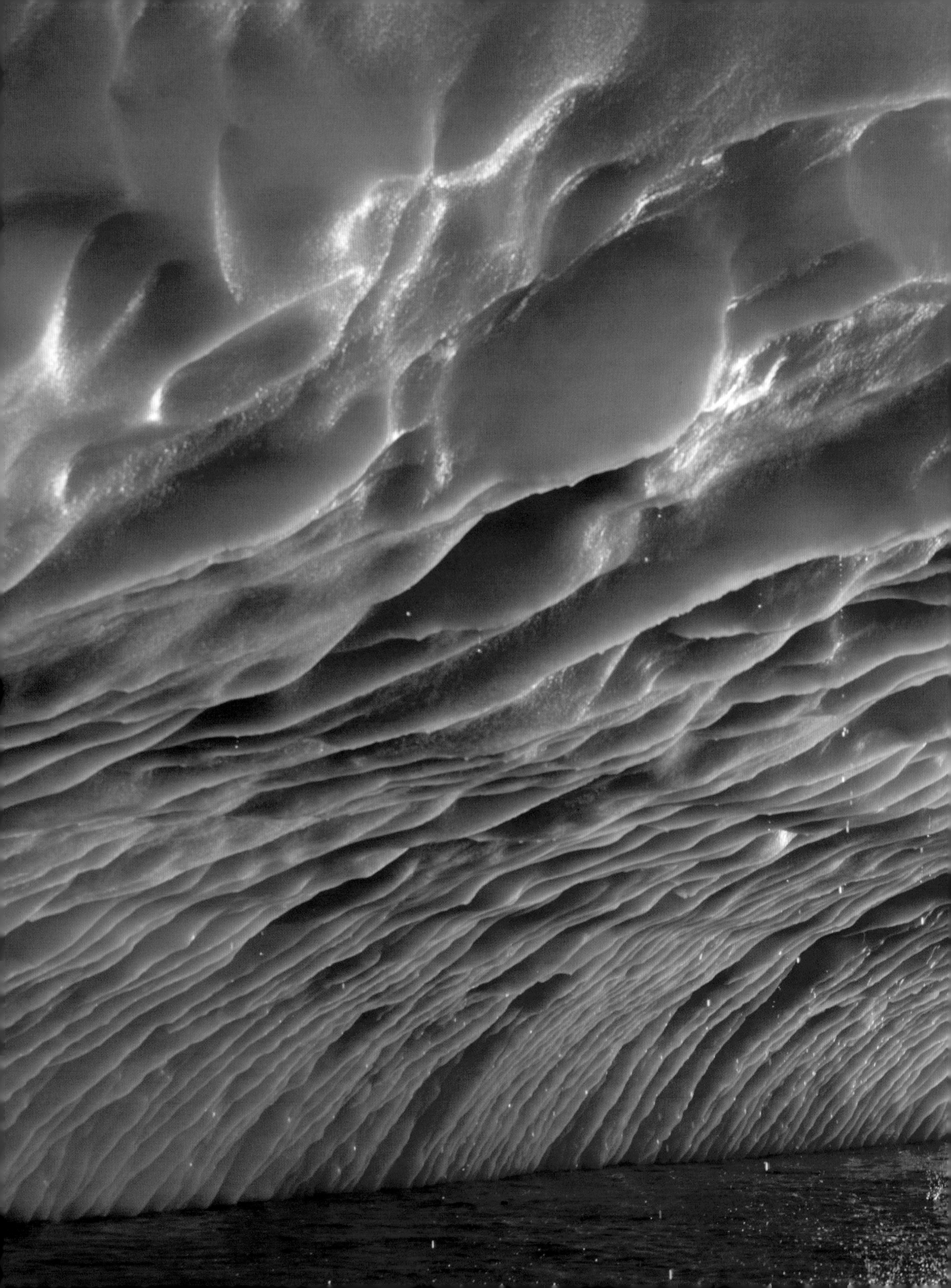

Since its very earliest years, Earth has switched back and forth between two kinds of world: one that is completely ice-free and enjoys a balmy greenhouse climate, and one in which ice covers a significant proportion of the planet. This seesaw between greenhouse and icehouse seems to be driven by the interplay of several processes, including rhythmic variations in Earth's orbit and the ever-changing geometry of continents and oceans. Rather like intermeshing biorhythms, planetary events sometimes conspire to ensure that Earth is distant from the Sun at a time when landmasses are concentrated at the poles, and the world slides inexorably into an ice age. Severe cold snaps are rare for Earth – the planet has been ice-free for nearly 90 per cent its existence – but they have happened at least four times, and as we saw in chapter three, some of them seemed to be linked to great leaps forward in the evolution of life. The most recent glacial epoch began 1.8 million years ago and we are still in the thick of it, with vast sheets of ice almost completely covering Greenland and Antarctica. Although the glaciers have waxed and waned, our species has never known an ice-free world, and Earth's changeable glacial climate has shaped our evolution; we are creatures of the ice age.

From Snow to Ice

The vast bulk of the ice on our planet began life as snow (*see* 'A Letter from the Sky', page 182). The ice crystals that grow into snowflakes can form anywhere in Earth's atmosphere. Even in the tropics there is snow in the sky, confined to high clouds where the temperature is well below freezing. But only where freezing conditions extend to ground level can snow reach the surface, and only in the coldest places does it linger for any length of time. In polar regions and at high altitudes more snow falls each year than can melt, and the excess steadily builds up. It partially melts and then freezes again, compacting and recrystallizing into small round grains of ice called firn. New layers of snow fall on top, and their growing weight squeezes the firn grains ever more tightly together, fusing them. By the time the ice is several tens of metres deep, the firn has been transformed into interlocking crystals of ice the size of footballs. It becomes even denser still, gradually turning into glacier ice and, crushed by the overwhelming weight of ice above, it begins to flow downhill.

Depending on how much melting and recrystallization goes on at the surface, the conversion of snow to glacier ice can take anything from a few years to hundreds. But in the end the ice succumbs to gravity and begins to move. Indeed, for all its static, frozen appearance, all the ice on the planet is in motion. The first Englishman to lay eyes on the spectacular Mer de Glace glacier at Chamonix in the French Alps described it as 'an agitated sea that seemed suddenly to have become frozen'. Hence its French name, which means 'sea of ice' – a description that most expressively captures the nature of the world's frozen wastes.

Earth has about 30 million cubic km (7 million cubic miles) of ice – enough to cover the entire planet to a depth of 60 metres (200 feet) if it were spread out evenly. Fortunately for us, it is concentrated in only certain parts of the globe, such as the great ice sheets that sprawl across Antarctica and Greenland or the smaller glaciers that flow from snowcapped mountains. Few of us are lucky enough to witness the awe-inspiring sight, sound and feel of a volcanic eruption, but to stand beside a glacier is easily the next best thing – and it is certainly a safer and more reliable way to experience the tremendous power of our planet's natural forces. The leading edge of a glacier is deceptive. Look at it and there is something faintly tawdry about it. The ice is not clear, sparkling or transparent – it is dirty, mixed with the sediment carried inside it and the earth and rock that is pushed along in front of it. In summer it can be slushy, melting and dripping to create pools of dirty water at your feet. The overwhelming colour is a bluish

OPPOSITE **THE LEADING EDGE OF THE ANTLER GLACIER, JUNEAU ICE FIELD, ALASKA, USA.**

A Letter from the Sky

In January 1885, US farmer Wilson Bentley took the world's first photograph of a snowflake. Bentley had been captivated by the exquisite beauty of snowflakes since childhood and had discovered a way of catching them on black velvet so that he could scrutinize their flower-like patterns more closely. But they melted too quickly to draw, so, with much trial and error, Bentley devised a way of connecting a bellows camera to a microscope and began taking pictures. He went on to take more than 5000, many of which were published in magazines, books and journals the world over. He called his subjects 'miracles of beauty' and made the famous observation that no two snowflakes are alike – a claim that has neither been proved nor disproved.

Snowflakes are simply ice crystals, and they can develop into a seemingly infinite variety of shapes because each one is sculpted by a unique set of atmospheric conditions. They begin life as 'seeds' – microscopic specks of airborne dust that provide a surface where water can crystallize out of the freezing air. Depending on the precise temperature and humidity, ice crystals can take on a range of different forms, including thin needles, hollow columns, hexagonal plates and six-sided stars. All have a hexagonal crystal structure that stems from the fundamental molecular structure of water, and so snowflakes usually have six sides. As the embryonic snowflake grows, crystals extend outwards from its core in six directions, branching and changing in form as atmospheric conditions vary around them. As it grows heavier the snowflake starts to fall, perhaps reaching a gentle 3–5 km/h (2–3 mph), and spins as it tumbles, which helps preserve its symmetry. Snowflakes are fragile and break easily, and in warmer air they become wet and stick together in irregular clumps. As a result, barely a quarter reach the ground with their beautiful geometrical structure intact. Even so, they can occasionally grow to spectacular proportions. Some of the largest on record fell in England in April 1951 and measured 12.5 cm (5 inches) across.

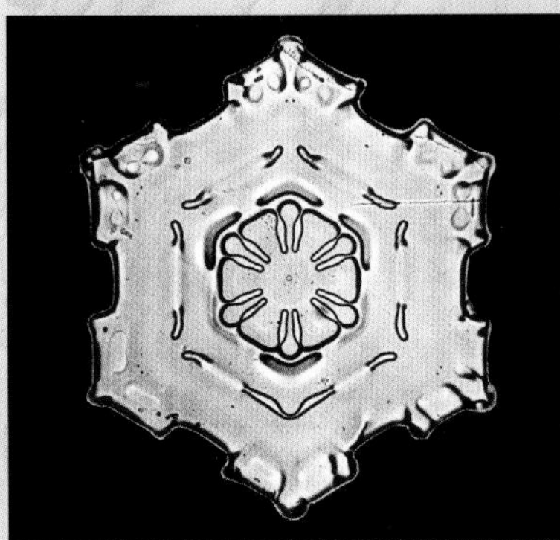

Bentley's photographs inspired Ukichiro Nakaya, a Japanese physicist, to embark on an ambitious programme of research into the structure of snow crystals. Nakaya spent many years photographing, categorizing and painstakingly re-creating snow crystals in his laboratory. He worked out the particular atmospheric conditions in which each type of crystal forms, and he summarized his work in a chart now known as the Nakaya diagram. With this, it is possible to read the meteorological story written in a snowflake and so deduce what atmospheric conditions are like high above. Nakaya referred to the snowflake as 'a letter from the sky'.

WILSON BENTLEY'S MICRO-PHOTOGRAPHS OF SNOWFLAKES REMAIN A DEFINITIVE SET OF IMAGES, SUGGESTING EACH CRYSTAL IS UNIQUE.

tinge of dirty grey, with a white coating trying hard to remain pure. And it seems so static to look at. But it is the sound of ice that strikes you. Not the quiet 'tink' of an ice cube gently bobbing in a cocktail glass, but something altogether more powerful and terrifying. Get close to the edge of a glacier, be it in an Alpine ski resort or the desolate landscape of Iceland, and what fills your ears are relentless, slow, deep sounds of groaning, breaking, creaking, like the ancient timbers of some giant square-rigger rolling at sea. Sudden cracking noises ring out at random, as though huge slabs of rock have finally given way under unimaginable pressure. For the ice is on the move, pushing solid rock out of its path and leaving a scoured and shattered landscape in its wake.

Rivers of Ice

Viewed from satellites, glaciers make an extraordinarily beautiful sight, their fluid nature clearly visible in the sinuous shapes they create as they ooze slowly downhill like giant rivers, swirling around mountains and weaving through valleys to spread out across frozen plains below, tearing at the land as they go. Even in the seemingly flat and desolate ice-covered plains of Antarctica, satellite images reveal undulating patterns reminiscent of ocean waves. For decades, explorers and scientists hiked across the ice sheets without realizing they were walking up and down over 'megadunes' of snow many kilometres long. One sea of dunes in East Antarctica covers an area the size of California, its gigantic ripples visible only from space.

But for the most dramatic view of moving ice, you need to look at a glacier not from above but from below. In Norway, the Svartisen icecap is drained by two glaciers, one of which is called Engabreen. It is not the largest glacier of its kind, nor the most fast moving. What makes it special is that the Norwegian Water

RIGHT **THE MEADE GLACIER IN ALASKA, USA, REVEALS THE FLUID NATURE OF THESE RIVERS OF ICE.**

Resources and Energy Directorate has excavated a unique experimental laboratory directly beneath it.

A flight, a long drive, a boat trip across a fjord, and a hike up a mountain take you to the entrance of the Svartisen Subglacial Laboratory, well inside the Arctic Circle. Engineers have drilled out a 1.6 km (1 mile) long rock tunnel from the base of the glacier, where scientists carry out their experiments on the ice. Far above them, the surface of the glacier appears static to the naked eye, its broken, fissured surface offering the only clue to the turmoil it is undergoing. Deep below, under 200 metres (650 feet) of ice, the scientists must constantly re-excavate their 10 metre (33 foot) long cave with high-pressure jets of hot water, because the ice is moving. Left unchecked, the cave would fill within a couple of days and become just another part of the grinding, sliding mass of ice. Water may be transparent, but deep under the glacier the cave is pitch black. Switch on a powerful light and the walls and ceiling of the heaving laboratory shine an astounding blue. Every now and then, large 'bubbles' of water that are held trapped within the ice burst open and shower the scientists with icy droplets. These bubbles of liquid water came as a surprise to the scientists, who as yet have no explanation for their abundance. What is clear, however, is that the ice is much more plastic than had been thought – it deforms easily, like toothpaste, pressing into every space that becomes available. The experiments may shed more light on the glacier's curious mechanics as the scientists gather data about its speed, temperature, and the enormous stresses it exerts at its interface with the underlying rock. At the base of the glacier, the ice is truly dirty, mixed up with grit, gravel and broken rocks, all of which grind and scrape the land below like sandpaper on wood. It is this part of a glacier that shapes the planet.

LEFT **AN ICE CAVE UNDER MUIR GLACIER IN ALASKA'S GLACIER BAY NATIONAL PARK. THESE EPHEMERAL AND UNSTABLE HOLLOWS CAN FORM NATURALLY AS MELTWATER RUNS UNDER THE GLACIER.**

Valleys eroded by rivers are typically V-shaped. Flowing water seeks out the lowest ground and cuts an ever deeper channel into it, while the ground on either side erodes more slowly to form slopes. Not so a glacier. Ice fills the valley and moves through it like a bulldozer, scouring every surface with rocky debris. Vast amounts of sediment and rubble are swept along, sometimes for many kilometres, finally to be dumped in sodden heaps in the 'ablation zone', where the glacier's leading edge is melting. Because the glacier erodes the valley's walls as well as its floor, it carves out a characteristic U-shape. Such U-shaped valleys are not just to be found among mountains – they can be seen all over the northern hemisphere, a legacy of the great glaciers and ice sheets of the ice age.

Ice-age glaciers left behind many other telltale signs of their presence. Piles of dumped rubble, called moraines, today form distinctive mounds and ridges. More streamlined heaps of debris, carved into teardrop shapes by glaciers sliding over them, are known as drumlins. One of the world's largest drumlin fields, consisting of some 10,000 teardrop-shaped hills, can be found to the east of Rochester in New York State. The huge boulders in New York's Central Park are 'erratics' carried by glaciers from New Jersey, and Coney Island is part of the plain of sand and gravel that washed out from a melting glacier. Even the skyline of Manhattan owes its shape to the ice sheet that once covered the island. The bedrock was eroded more deeply in the centre of Manhattan than at the ends, and deposits of clay were left in the glacier's wake. The exposed bedrock at the northern and southern ends of the island was firm enough to support the foundations of the tallest buildings, while the clays left in the middle were not. And as for Long Island to the east, that is a 'terminal moraine' – a heap of debris dumped at the end of a glacier.

RIGHT **THESE ERRATICS WERE LEFT BEHIND BY THE RETREATING GLACIER THAT POLISHED THE SMOOTH ROCK SURFACE OF YOSEMITE NATIONAL PARK IN THE USA.**

Secrets of the Ice

In the summer of 1942 the shortage of warplanes stationed in Britain was acute, and US military leaders initiated Operation Bolero in an effort to build up their strategic forces. Carrying planes by ship was risky because German U-boats were targeting Allied convoys, so it was decided that some of the aircraft should be flown to the UK via refuelling points in Labrador, Greenland and Iceland. On 15 July, 25 US Army Air Corps men were on just such a mission, delivering two B-17 Flying Fortress bombers and six P-38 Lightning fighters, when they ran into appalling weather over Greenland. They were forced to land on the ice and abandon the aircraft. Snowed on year after year, the planes slowly sank from view until they gradually became a part of Greenland's icecap.

Forty-six years later, a team of enthusiasts returned to locate and recover the lost squadron, using ground-penetrating radar to pinpoint the planes' location. Snowfall records suggested they should be about 15 metres (50 feet) deep, but they turned out to be under 80 metres (260 feet) of glacier ice. Half a century of glacial build-up had not served the aircraft well. An attempt to dig out one of the Flying Fortresses revealed that it had suffered serious structural damage due to the constant distortion of the ice in which it was entombed. A later attempt to recover one of the Lightning P-38s was successful, the smaller planes having proved more resistant to the crushing forces within the ice, and the fighter has now been restored and flown to the UK to complete its delivery – just 65 years too late. The lost squadron was not only of interest to aviation enthusiasts – it also provided remarkable evidence of the rate at which Greenland's ice accumulates, which is of critical interest to scientists investigating climate change.

Even more astonishing things have been found hidden in ice. Perhaps the most famous discovery was 'Otzi', the mummified Stone Age hunter who died on a high Alpine pass 5000 years ago and remained entombed until climbers found him in 1991. His body and possessions were so well preserved that the climbers thought he must be a modern corpse. Whole mammoth carcasses have been found in the frozen wastes of Siberia, prompting speculation that the species could be brought back to life. Or there is the case of *Stardust*, a British airliner that vanished while crossing the Andes in 1947. No remains were found until 50 years later, when its wreckage and some of the victims' bodies emerged from a glacier that had carried them down the mountain.

But it is only in recent years that the full extent of what ice can hide has been realized. In 1996, ground-penetrating radar, seismic surveys and satellite images revealed that a vast lake of liquid water lay hidden 4 km (2.5 miles) deep under Antarctica's main ice sheet. Dubbed Lake Vostok, it has been sealed in ice for at least half a million years and possibly much longer. In 1998, scientists drilled into the ice sheet and took samples from close to the top of the lake, taking care not to break through to the water. The samples contained microbial life, hinting that the lake may harbour a unique ecosystem that has evolved in isolation from the rest of the planet. But to study the lake itself presents scientists with a real dilemma: they cannot sample the water without breaking through the ice, ending the lake's long isolation and almost certainly contaminating its pristine, prehistoric waters.

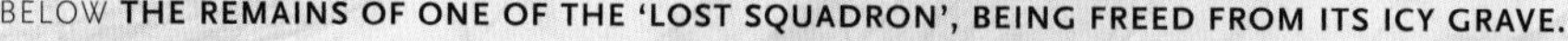

BELOW **THE REMAINS OF ONE OF THE 'LOST SQUADRON', BEING FREED FROM ITS ICY GRAVE.**

ABOVE **MANHATTAN'S DISTINCTIVE SKYLINE IS A LEGACY OF THE ICE AGE. THE TALLEST BUILDINGS ARE CLUSTERED AT THE ENDS OF THE ISLAND, WHERE THE UNDERLYING BEDROCK SUFFERED THE LEAST EROSION BY GLACIERS.**

Ice-age features are much easier to see in the stunning scenery of Yosemite National Park in California. The park's centrepiece – Yosemite Valley – is one of the world's finest examples of a U-shaped valley. The remarkable sheer granite cliffs of El Capitan and Half Dome were carved out and then polished by ice. As the glacier swept down the valley, ice tore into its sides, undercutting them and causing the rock to fracture and fragment. Only the hardest rock survived to become the majestic cliffs that visitors now flock to in their thousands. Look carefully and you can see horizontal scratch marks in the rock where debris scraped across it, surrounded by an otherwise smoothly polished surface. Throughout Yosemite you can also find spectacular boulders balanced on narrow pedestals: erratics that remained intact while the softer rock on to which the glacier dropped them eroded away to a narrow pinnacle of support. Far above the main valley floor are smaller, adjoining U-shaped valleys left high and dry when the ice retreated, their small glaciers having cut much less deeply into the ground than the main glacier they were feeding. Today, waterfalls like Bridalveil Falls pour out of these 'hanging valleys' in summer and cascade down the glittering granite walls, adding to the spectacular view from below.

Snowball Earth

As we saw in chapter two, Earth seems to maintain a surface temperature just right for life through the workings of its global thermostat. When things get too warm, the rate of weathering increases and so carbon dioxide is removed from the air by rain, bringing the temperature

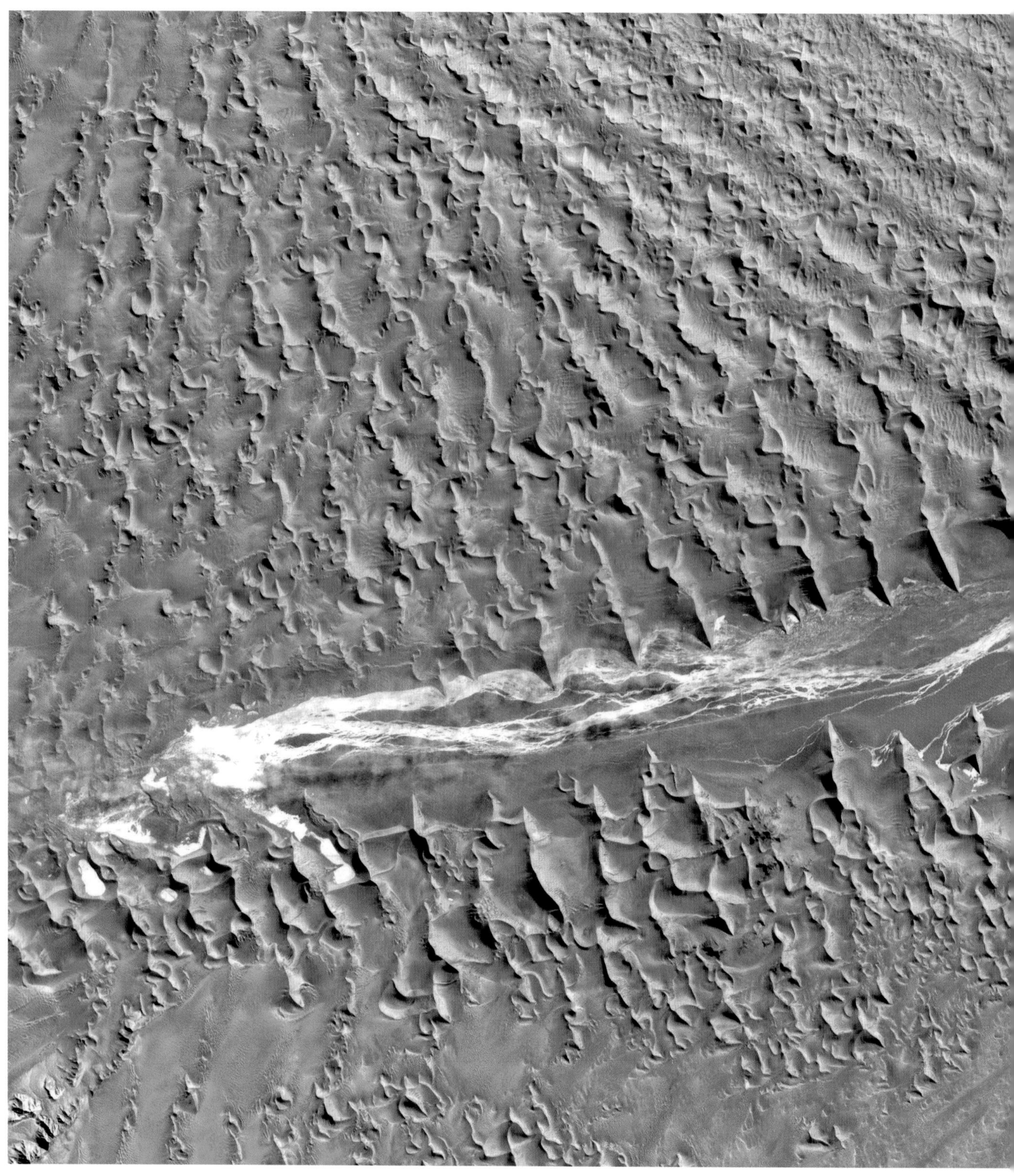

back down; when things get too cool, the rate of weathering falls and carbon dioxide builds up again, making the planet warmer. Over billions of years, this self-correcting system has been amazingly effective. But it isn't perfect. Every so often it seems to have broken down, making the planet's climate go haywire.

For more than half a century, geologists have been puzzled by glacial formations in parts of the world that don't make sense. In Namibia, for example, there are spectacular examples of 'drop stones' – glacial boulders embedded in sedimentary rock that formed around them after they were dumped by glaciers. The trouble is that Namibia was nowhere near the poles when these rocks were dropped. In fact, it was even closer to the equator than it is now. The inescapable conclusion is that ice must have covered far more of the planet than it does today, perhaps even reaching all the way to the equator to create an icebound world: Snowball Earth.

This seems to have happened several times. In chapter three we saw that a Snowball Earth event may have taken place around 2.45 billion years ago when oxygen levels were rising, but it isn't clear whether ice reached all the way to the equator at that time. Much more compelling evidence of a Snowball Earth comes from the period between 780 and 630 million years ago, when there appear to have been at least two freezing phases when ice encased the entire globe. No-one knows for sure what tipped the climate out of balance, but Earth's ability to stabilize its own temperature clearly suffered a catastrophic breakdown. As the planet cooled, vast sheets of ice would have spread out from the poles. The advancing ice reflected ever more of the Sun's heat back out to space, making the world even colder, which led to more ice, and so on. Earth was trapped in a runaway deep freeze.

LEFT **TODAY NAMIBIA'S SUN-SCORCHED LANDSCAPE IS BURIED UNDER A SEA OF GIGANTIC SAND DUNES, BUT EVIDENCE FROM ROCKS IN THE REGION SUGGEST THE LAND WAS ONCE COVERED IN ICE.**

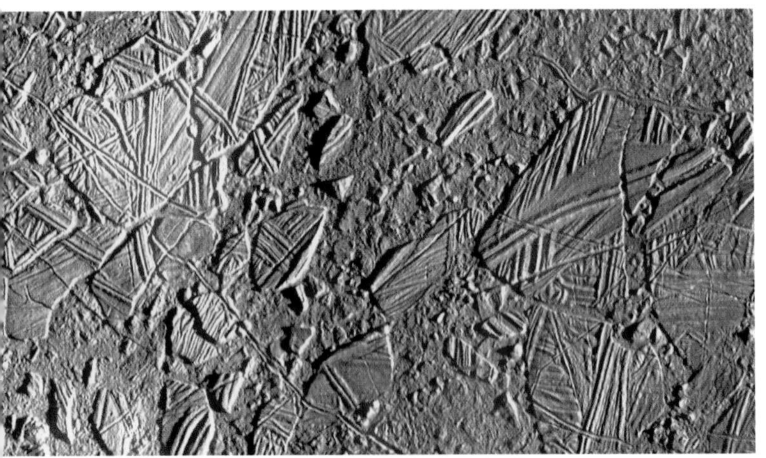

It's hard to imagine just how horrendous the conditions were, but Earth must have been a truly harsh place during its coldest years. Today the sea ice that covers the Arctic Ocean is some 2 metres (7 feet) thick; during the Snowball era it would have been up to 1000 metres (3300 feet) thick. Today Earth's average surface temperature is 15°C (59°F); during the Snowball era it would have been −50°C (−58°F), and even at the equator the temperature would have averaged −20°C (−4°F). With ice covering the ocean, air and sea could no longer exchange gas and moisture. There could be no evaporation and therefore no rain. Deep in the oceans, the water would have become toxic as chemicals spewed out by volcanic vents accumulated. The atmosphere would have been bone dry, and since there was no rain or snow, the continental interior probably became an arid desert like the 'Dry Valleys' of modern Antarctica, where ice evaporates directly into the parched air without first turning to water. Howling winds would have scoured the land, whipping up dust and flinging it over the ice sheets. Perhaps Snowball Earth wasn't dazzling white at all. Viewed from space, the planet's face probably had a sickly, mottled complexion, dirty white with smears of brown and grey.

The frightening thought is that our planet could have been trapped in this freezer for ever. For once ice had completely entombed the planet, so much of the Sun's heat would have been reflected away that Earth might never have been able to warm up again. But in the end, the planet's thermostat did recover. The world finally emerged from under the ice some 635 million years ago, saved by greenhouse gas emissions from volcanoes, and as we saw in chapter three, the thaw was followed by an incredible flourishing of life. The first animals appeared, and soon after they exploded into a spectacular diversity of forms. It was in this riotous burst of evolution, known as the Cambrian explosion, that the fundamental body plans of all creatures alive today were laid down.

Megaflood

Ice can create landscapes of breathtaking natural beauty as it slowly scours Earth's surface, but the effects of melting ice can be even more dramatic. In October 1996, Iceland's Grímsvötn volcano – which lies beneath the largest glacier in Europe, Vatnajökull – began to erupt. Clouds of steam rose from beneath the ice, and the centre of the glacier began to sag ominously. The bottom of the glacier was melting, and water was filling the huge volcanic crater. The eruptions ceased but the subglacial lake continued to grow for another three weeks, swelling in volume to nearly 4 cubic km (1 cubic mile), until, on 5 November, an earthquake triggered the inevitable glacial outburst, known in Iceland as a jökulhlaup. Water swept from beneath the glacier at a rate of some 50,000 cubic metres (1.8 million cubic feet) a second, destroying roads, bridges and power lines. And in a final twist, the volcano itself re-erupted the following day, an event almost certainly triggered by the release of pressure as the great weight of water flooded away.

The Grímsvötn outburst flood was huge by today's

ABOVE **THE SURFACE OF JUPITER'S ICEBOUND MOON EUROPA IS CRAZED WITH FRACTURES THAT HAVE SPLIT AND REFROZEN. OF ALL THE SOLAR SYSTEM'S WORLDS, EUROPA IS PROBABLY THE CLOSEST IN APPEARANCE TO SNOWBALL EARTH.** OPPOSITE **THE STEEP WALLS OF THIS U-SHAPED VALLEY IN WYOMING, USA, ARE SURE SIGNS THAT IT WAS CARVED BY A GLACIER.**

standards, but in comparison with outburst floods of the distant past it was minuscule. In Washington State, USA, a vast area of land, some 40,000 square km (15,000 square miles) in area, is etched with peculiar channels, potholes and ripple marks quite unlike anything else on Earth. The area is known as the Channelled Scablands, and its unique erosion features are the scars of what can only be described as a megaflood. Towards the end of the last ice age, an immense ice sheet (the Cordilleran ice sheet) covered much of western North America. Its southern reaches petered out among the Rocky Mountains, where glacial lobes extended like fingers across highlands and valleys, blocking meltwater rivers with dams of ice. Huge lakes built up behind these dams. The largest was Lake Missoula, which at its greatest extent held more water than today's Lake Ontario and Lake Erie combined. At 300 km (190 miles) wide, it was effectively an inland sea. Missoula's water level rose until the lake was some 600 metres (2000 feet) deep, but then the glacier at the base of the ice dam became buoyant and began to lift away from the rocky bed. The resulting outburst flood was almost incomprehensibly vast in its scale and power. The lake drained in only two days, its 2000 cubic kilometres (500 cubic miles) of water pouring out at up to 100 km/h (62 mph) and tearing across Washington State towards the Pacific Ocean.

The first signs of the coming flood would have been a gale-force wind and shaking of the ground as a mass of compressed air preceded the towering wall of water. The roar would have begun half an hour before the flood-wall struck, with winds increasing in intensity until they were uprooting trees and creating a blinding dust storm, followed by torrential rain. Then came the flood. Every living creature would have been swept to

RIGHT **THE CHANNELLED SCABLANDS OF THE NORTHWEST USA WERE SCOURED OUT BY A SERIES OF GIGANTIC 'MEGAFLOODS' FROM AN ICE-AGE GLACIAL LAKE.**

its death and carried far downstream, where the bones of mammoths and other big herbivores have been found scattered. The noise would have been deafening, with boulders and icebergs smashing together, huge waves crashing down over violent, roiling rapids, and rocks, boulders and trees being flung high over the water. Powerful whirlpools captured boulders and ice and churned them around, while the spinning water drilled down into the ground, boring the world's largest potholes out of solid rock. All the soil was stripped off the land, exposing the rough basalt rock beneath, which in turn was scoured to form the Scablands. Some 40 cubic km (10 cubic miles) of water gushed out of the lake every hour – a rate far greater than the combined flow of all the rivers in the world today – and the sediment it washed from the land was swept 2000 km (1200 miles) out into the Pacific Ocean. In a matter of hours, the roaring waters gouged out deep canyons, channels and waterfalls, one of which was 150 metres (490 feet) tall and 5 km (3 miles) wide – ten times the size of today's Niagara Falls. Today you can still see giant ripple marks taller than houses, where mountains of gravel were dumped in the wake of the flood, as well as 200-tonne boulders littering the landscape, some of them having been carried from as far as 1000 km (620 miles) away.

And Lake Missoula didn't flood just once. The ice dam quickly settled back into place after the lake emptied, allowing it to refill in a few decades – only to burst open again as the glacier became unstable once more. The landscape shows signs of 40 or more such dramatic outpourings, though they gradually weakened as the ice sheets shrank away at the end of the ice age. Catastrophic as they certainly must have been for anything caught in their path – including perhaps some of the very first people to reach continental America – these periodic megafloods were not entirely without merit. The sediment they laid down formed the fertile soils of western Oregon and, if nothing else, provided the *terroir* for some of the best vineyards in the world.

Floating Water

Ice has not only shaped Earth's landscapes – it has also shaped the course of evolution, and it has done so because of a very peculiar property. Water is unlike all other substances in one critical respect: it expands when it freezes. Most substances expand as they warm up, becoming less dense and lighter as they do so. So it is for water in its liquid and gaseous states: warm it and it expands and becomes less dense; cool it and it contracts and becomes denser. As we saw in the previous chapter, it is the changing density of water cooling down and warming up that drives the ocean's global conveyor. So far so good. But as water cools below 4°C (39°F) and nears its freezing point, its molecules begin to take on the more open, crystalline configuration of ice, making the water expand and become lighter. The result is that solid water floats on liquid water – something that almost no other substance is known to do.

Imagine a world in which the opposite happened. Instead of freezing across the top, lakes, ponds and rivers would freeze from the bottom up, with crystals of ice falling from the air-chilled surface like snow. Without a layer of surface ice to insulate deeper water from freezing air, bodies of water would freeze in their entirety. As it is, lakes and ponds nearly always have a layer of liquid water at the bottom, even when they appear to be frozen solid. And because water is at its densest and heaviest at 4°C (39°F), the water at the bottom maintains this temperature constantly, providing a stable environment in which living things can tough out the worst of winters. Even during the Snowball Earth era, when glaciers smothered the whole planet, the remarkable property of water ensured that havens for life remained.

The ability of ice to float is responsible for one of the natural world's greatest spectacles: iceberg nurseries. In central Greenland, the fresh snow falling on

OVERLEAF **A SHIP IS DWARFED BY ICEBERGS OFF GRAHAM-LAND PENINSULA, ANTARCTICA.**

Noah's Flood

One controversial theory says that during the last ice age, when sea levels were at least 100 metres (330 feet) lower than today, the Mediterranean Sea and the Black Sea were separated by a strip of land across the site of today's Bosphorus channel. Cut off from the ocean, the Black Sea would have been no more than a low-lying lake. Evidence for the theory comes from sediment samples drilled from the bottom of the Black Sea that contain mud and shellfish normally found in river deltas, suggesting that the sea was much smaller and shallower than today, fed only by the rivers flowing from central Europe. Divers have also found what seem to be traces of human settlements in the depths of the sea, where there may once have been land.

The theory is that the Mediterranean slowly rose after the ice age ended until, around 7500 years ago, the rising waters breached the land bridge and poured over the top in a raging torrent 200 times bigger than Niagara Falls. And so the arid basin around the lake filled up to become t he Black Sea. It may sound far-fetched, but a similar event may have occurred 5 million years ago when the Atlantic Ocean broke through the Strait of Gibraltar to refill the Mediterranean, which had dried out to become a shrunken, salty vestige of its former self.

The two scientists who proposed the Black Sea flood theory say it could be the historical basis for the story of Noah's flood in the book of Genesis. Such a disaster, they say, would have driven coastal communities into neighbouring lands, taking the story with them, and perhaps also the agricultural know-how that fostered the earliest civilizations. But other scientists disagree, citing evidence that the flow between the Black Sea and the Mediterranean existed throughout the ice age. It remains to be seen who will turn out to be right.

BELOW **TODAY THE MEDITERRANEAN AND THE BLACK SEA ARE JOINED BY THE BOSPHORUS STRAIT, BUT THE NARROW STRIP OF LAND BETWEEN THEM MAY ONCE HAVE FORMED AN UNBROKEN BARRIER, UNTIL RISING SEA LEVELS BREACHED THE DAM.**

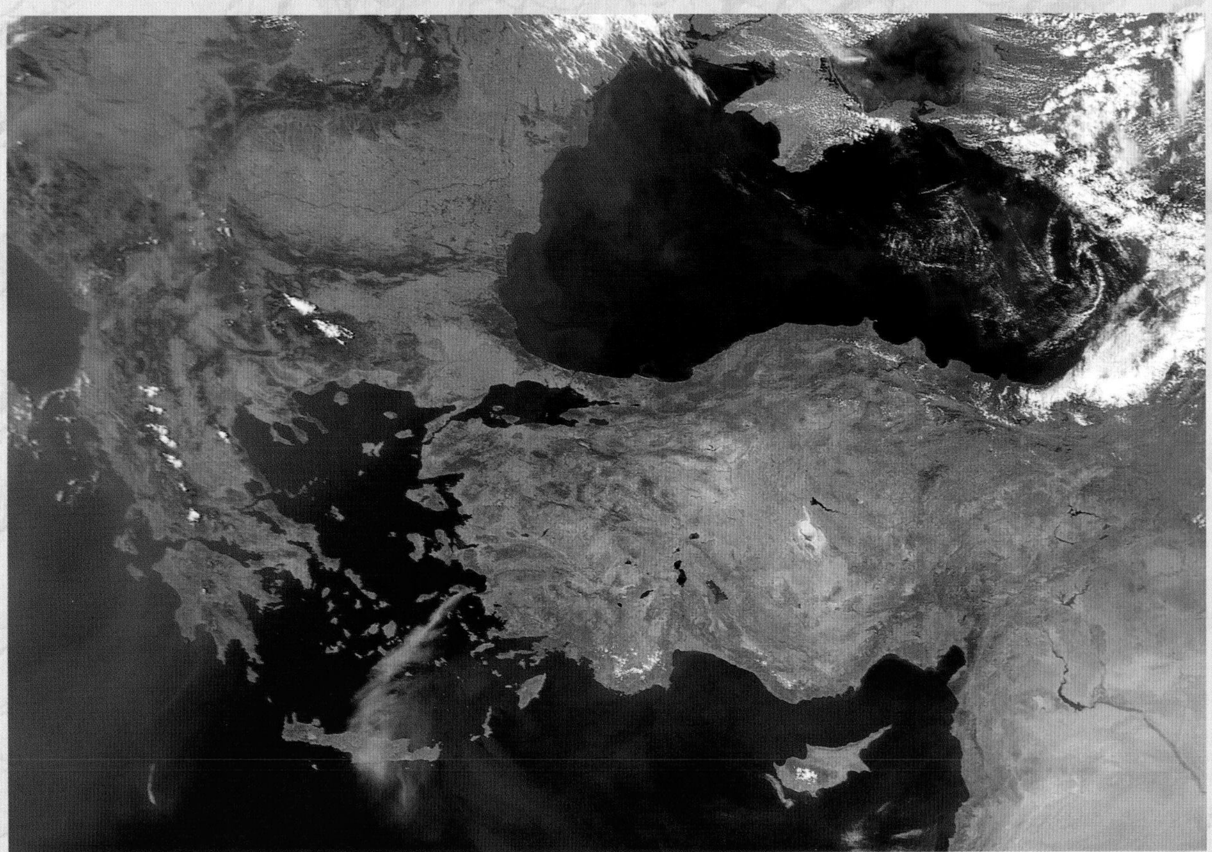

the ice sheet marks the beginning of a 100,000-year journey that the ice will make as it flows through a glacial conveyor belt towards the coast, finally to be expelled as icebergs. Greenland's ice sheet is the largest mass of ice on the planet, after the great ice sheets of Antarctica. Its vital statistics defy belief. It is more than 1.7 million square km (650,000 square miles) in area and contains 2.5 million cubic km (0.6 million cubic miles) of ice – 10 per cent of the world's total. Its average thickness is 1500 metres (5000 feet), rising to 3200 metres (10,500 feet) at the peak of the 'southern dome', and its monumental weight has crushed the land below, pushing the underlying bedrock down to sea level. If it were to melt, along with the other smaller icecaps in the Arctic area, global sea levels would rise by 6–7 metres (20–23 feet).

It is far from being a solid mass. Ice accumulates where snow falls in Greenland's higher central regions and then flows outwards, feeding coastal glaciers that slide inexorably towards the sea. Where the glaciers terminate, immense blocks of ice splinter away from the main mass and collapse thunderously into the water to become icebergs – a process known as 'calving'. Greenland's fastest moving glacier is Jakobshavn Isbræ on the west coast. It flows into the ocean at a rate of 30 metres (100 feet) a day, disgorging some 35 billion tonnes of icebergs annually – a tenth of Greenland's total output.

From the main Greenland ice sheet, Jakobshavn Isbræ extends like a long finger, stretching some 40 km (25 miles) west through a narrow 'ice fjord' to reach the sea. On the north side of the glacier's mouth sits the small town of Ilulissat, Greenland's third largest settlement, with a population of 4500. Its Greenlandic name means simply 'the icebergs'. Around 50 cubic km (12 cubic miles) of icebergs flow past this modest town

RIGHT **IN THIS SATELLITE IMAGE ICEBERGS ARE CLEARLY SEEN FRAGMENTING FROM THE LEADING EDGE (BOTTOM LEFT) OF EUGENIE GLACIER IN THE CANADIAN ARCTIC.**

each year. Grinding, cracking and crashing together, the vast tower blocks shove their way out of the fjord's narrow mouth – which has earned the nickname 'iceberg alley' – and scatter into the ocean. It is an astonishing, humbling sight. The peaks appear to float around each other in a majestic dance, but under the water, where their vast bulk lies hidden, the icy monsters are clashing, jostling and colliding with each other as they speed out of the fjord. Some are so enormous that they become stuck to the sea bed until the weight of traffic behind them forces them onwards.

After escaping from the fjord, the icebergs set off on a journey of decay. Ocean currents carry them first north past Ilulissat and then sweep them round to the south, taking them across the Labrador Sea and out into the North Atlantic, where they cross transatlantic shipping lanes. Some of the larger icebergs drift even further south than the UK or New York City, before finally melting and diluting the salty waters of the North Atlantic. As we saw in the previous chapter, an increase in the flow of fresh water into the North Atlantic could interfere with global ocean currents, so it is all the more alarming that things seem to be changing in iceberg alley. Scientists have been surveying the Jakobshavn Isbræ glacier by satellite since 1991, and their results show that its rate of flow has doubled in the last decade. The increased output has raised the rate at which global sea levels are rising by 0.06 mm (0.002 inches) a year, and an extra 25 cubic km (6 cubic miles) of fresh water is diluting the ocean every year. The consequences are yet to become clear.

Cycles of Cold

Greenland's ice sheet contains more than just ice. When snow lands, air gets caught in the spaces between the snowflakes, and some of it ends up trapped in tiny bubbles as the snow turns slowly into glacier ice. In effect, these bubbles are samples of the atmosphere that gradually work their way deeper into the ice sheet, sinking over hundreds and thousands of years. By drilling out a long section, or 'core', of ice and retrieving the bubbles trapped within it, scientists can see how the atmosphere has changed over time. These ice cores not only yield fascinating information about past levels of gases such as oxygen and carbon dioxide in the air, they can also tell us how the Earth's temperature has changed.

Clues about temperature come from the oxygen atoms in water molecules. In essence, a tiny proportion of the oxygen atoms have an extra two neutrons in their atomic nuclei, making them heavier. These variant atoms, or 'isotopes', are known as oxygen-18, while the normal, lighter atoms are oxygen-16. Being heavier, water molecules containing oxygen-18 need more heat energy to make them evaporate, and they condense more easily too. Because of this difference, the ratio of the two isotopes in ice can be read like a thermometer. During warm periods, the amount of oxygen-18 evaporating from the sea increases, raising the ratio in the water that falls as rain and snow. The changing ratio in ice cores is so clear that scientists can even see peaks and troughs corresponding to summers and winters, and by counting back through these they can figure out the age of the ice. (A similar technique, based on the oxygen isotopes locked up in the skeletal remains of sea creatures in ocean sediments, can reach even further back in time than ice cores.)

In Greenland, ice cores have been drilled right down to the bedrock, and layers counted back through 110,000 years. At the Russian Vostok Station in Antarctica, the ice sheet was drilled to a depth of 3350 metres (11,000 feet), providing an atmospheric timeline reaching back 426,000 years. Most recently, a European project at Concordia Station in Antarctica drilled down through 900,000 years of ice. Thanks to such projects, scientists have constructed a detailed record of Earth's climate history, making it possible to track the cooling effect of volcanic eruptions or correlate carbon dioxide levels with global temperature; amazingly, Antarctic cores have shown that carbon dioxide levels are currently the highest they've been in

ABOVE **THE GREENLAND ICECAP HOLDS 10 PER CENT OF THE WORLD'S ICE. ITS WEIGHT HAS COMPRESSED THE ROCK BELOW DOWN TO SEA LEVEL.**

the last 800,000 years. And a wonderfully clear picture is emerging of the climatic oscillations that accompanied our planet's most recent ice ages.

Throughout the nineteenth century, as the science of geology advanced, telltale signs of glaciers were found in ice-free landscapes all over the northern hemisphere. It became clear that glaciers had advanced and retreated repeatedly in the past – there had been not one ice age but many – but exactly why remained a mystery. In 1842 a French mathematician, Joseph Alphonse Adhemar, suggested that periodic changes in Earth's orbit around the Sun might offer an explanation, but Adhemar didn't work out his theory in full, and the idea remained controversial. It wasn't until the 1930s that the complicated maths behind Adhemar's idea was done properly, by Serbian civil engineer Milutin Milankoviç. Milankoviç discovered that Earth 'wobbles' in its orbit in a way that results in a 100,000-year cycle in the amount of heat reaching us from the Sun, corresponding neatly with the 100,000-year cycle of ice ages.

Milankoviç's cycle is caused by a combination of factors. First, Earth's orbit is not truly circular. It is an ellipse, and it is not even a consistent ellipse, but varies from being nearly circular to being much more elliptical (due to the varying gravitational pull of Jupiter and

ABOVE **IN THE 'LITTLE ICE AGE', FROST FAIRS WERE A NORMAL OCCURRENCE ON THE FROZEN RIVER THAMES, AS DEPICTED IN THIS 17TH-CENTURY PRINT.**
OPPOSITE **THE LARSEN B ICE SHELF IN ANTARCTICA BEGAN TO FRAGMENT AT THE TURN OF THE MILLENNIUM.**

Saturn). Right now, Earth is quite close to its most circular orbit. Second, the tilt of Earth's axis of rotation varies. At present, Earth tilts about 23.5° from the vertical, but the figure varies from 21.5° to 24.5° and back over a cycle of 41,000 years. Third, the axis of Earth's rotation itself rotates, turning around slowly like a leaning gyroscope every 26,000 years. (Since Milankoviç's death, a fourth factor has been discovered: the plane of Earth's orbit tilts back and fourth on a 70,000-year cycle.) When the sums are done, the net result of all this subtle variation is a cycle lasting 100,000 years.

Earth's most recent cold spell lasted from 70,000 to 10,000 years ago. This was 'the Ice Age', the period that conjures up images of Neanderthals, mammoths, sabre-toothed tigers, and the enigmatic cave paintings of Lascaux in southern France. But strictly speaking, we are still in an ice age. The bitterly cold era in which the Neanderthals perished was merely a particularly intense phase of it, a fluctuation brought on by the orbital rhythms that Milankoviç identified. Our current relatively warm period is just a temporary lull – an 'interglacial'.

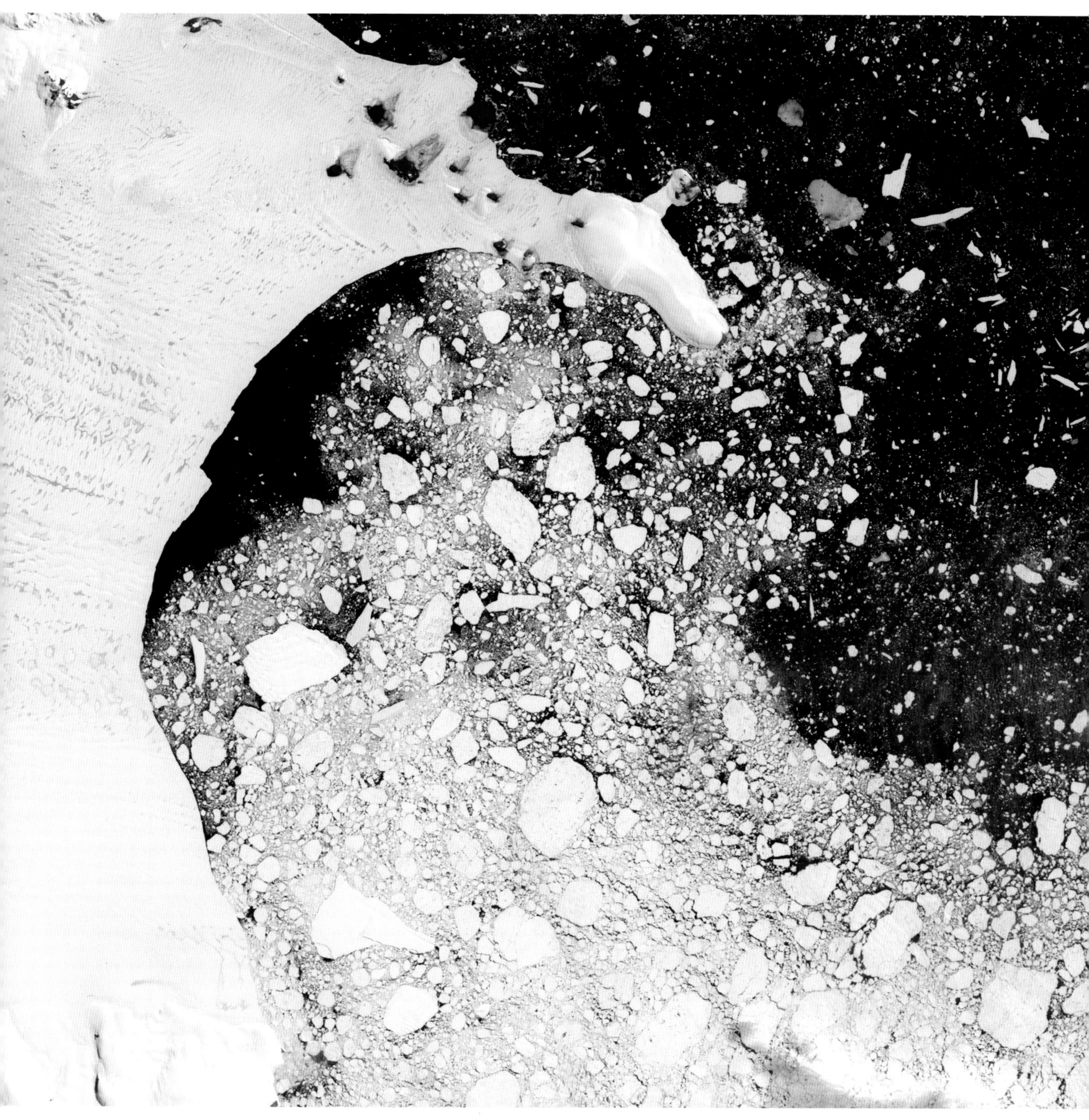

There have been at least four great ice ages in Earth's history, including the Snowball era. The most recent great ice age began around 40 million years ago, when ice sheets began to bury Antarctica. The last 3 million years have been the coldest, with ice sheets spreading across the northern hemisphere as well. For reasons no-one quite understands, Earth's climate also seems to have been unusually sensitive to its orbital variations in these 3 million years, with temperatures fluctuating up and down dramatically. And there is mounting evidence that these temperature swings can happen with alarming speed. In the previous chapter we saw how prehistoric beetles excavated from Cumbrian mud revealed that the northern hemisphere was plunged back into a mini ice age around 13,000 years ago when it was trying to warm up. Temperature readings from ice cores suggest the crash may have happened in as little as ten years, prompting speculation that the Gulf Stream must have shut down. But there is another reason why Earth's climate is prone to unpredictable lurches, and it lies with the ice itself.

When you look at a crescent Moon, the sharp white sickle stands out, lit by the Sun. But look closely and you may still be able to make out the Moon's whole disc, its dark side a faint grey against the black night sky. The faint grey light is 'Earthshine': sunlight reflected by Earth on to the dark side of the Moon. Every part of the Earth's surface reflects sunlight to some extent, but nowhere does so more effectively

ABOVE AND OPPOSITE **EARLY PHOTOGRAPHS HAVE BEEN USEFUL IN CHARTING THE RAPID RETREAT OF GLACIERS ACROSS THE NORTHERN HEMISPHERE, AS IN THESE BEFORE-AND-AFTER IMAGES OF THE PASTERZE GLACIER IN AUSTRIA IN 1875 (ABOVE) AND ITS SITE IN 2004 (OPPOSITE).**

than areas of snow and ice, which reflect 80–90 per cent of the energy that falls on them – including the invisible infrared rays that transmit heat. Because of this high reflectivity, or 'albedo', ice sheets can reinforce a change in Earth's climate. When the climate cools and ice sheets advance, their growing area reflects ever more heat energy back into space, cooling the climate further. The result is a positive feedback loop that causes temperatures to spiral ever lower. It works in the opposite direction too: if the climate warms a little, be it due to a volcanic eruption or an orbital wobble, ice sheets shrink and reflect less heat from the Sun, which makes the climate warmer, causing more melting, and so on.

This feedback loop may have played a key role in triggering Earth's great ice ages. Major glaciations, including the current one, seem to have happened at times when the arrangement of continents obstructed the flow of warm ocean currents from the tropics to the poles. The poles cooled, allowing ice sheets to form, and the growing expanse of ice then reflected the Sun's heat away, depriving the planet of warmth. Whether or not the theory is correct, scientists are certainly concerned that a change in the planet's ice cover could have dangerous consequences in a warming world. And as we shall see later, Earth's ice sheets and glaciers are changing with disturbing speed.

The waxing and waning of polar ice sheets over the last few million years had effects all over the planet. Indeed, our own origins and evolution may have been driven by ice ages. When ice sheets advanced, so much water became locked up at the poles that the climate dried out in the tropics. East Africa's climate became much more variable, putting great pressure on its

CHAPTER SIX

Rare Earth

Earth is a remarkable, ever-changing planet, full of extraordinary natural beauty. The wondrous world that we are so familiar with has been shaped by fundamental forces operating on scales that we humans can barely grasp: the stupendous impacts that put the planet together; the internal heat engine that drives the restless surface; the tenuous atmosphere that shields and fuels us; swirling oceans that carry warmth and nutrients around the globe; and the ice sheets that appear now and then to stir things up. Amidst these amazing cogs and wheels of our planet's machinery there is another fundamental force at work. It is a force that was born from Earth but has since forged a unique partnership with our planet. It is the true life force of our planet: life itself.

Life is what makes Earth really special. Not just any old life, of course, but advanced, intelligent, sentient life. It remains to be seen whether other planets in our solar system harbour simple, single-celled microbes similar to those that can tolerate the extremities of our planet, but complex life forms – multicellular creatures with tissues and organs – are undoubtedly unique to this corner of our cosmic neighbourhood. Given the inconceivably vast size of the universe, it seems reasonable to suppose that complex life must exist elsewhere too.

OPPOSITE **THE FAMILIAR VIEW OF PLANET EARTH. WHEN YOU CONSIDER WHAT IT HAS TAKEN TO MAKE THIS ROCK-COATED METAL BALL A HOME FOR LIFE, THE UNIVERSE BEGINS TO SEEM LIKE A VERY LONELY PLACE.**

Surely other Earths are lurking somewhere out there, teeming with biological riches. There are certainly lots of planets orbiting other stars, and there is no shortage of stars. Thirty years ago, the eminent astronomer Carl Sagan used a few galactic facts and figures to guesstimate that there might be as many as a million advanced civilizations in our Milky Way galaxy alone – one of more than 100 billion galaxies in the universe.

Mind you, we haven't found them yet, and it's not as though we haven't been looking. Ever since we humans started peering up at the stars, we've been searching the heavens for signs of intelligence. Of course, it's possible we haven't been looking for long enough or in the right places, but there's another possibility: perhaps there simply isn't anything out there. Geologist Peter Ward and astronomer Don Brownlee have a modern term for this age-old notion that we're alone in the universe: 'rare Earth hypothesis'. At the heart of their theory is the realization that it took a series of incredibly improbable flukes and strokes of good fortune to make Earth habitable. To appreciate how extraordinarily unusual our planet may be, it is worth recalling a few of the defining moments in its history.

How to Build a Habitable Planet

Earth was born in the right part of space. Any closer to the Sun, like Venus or Mercury, and you enter a hellish realm that is too hot to sustain liquid water. Any further away, like Mars, and it would be too cold for liquid water. There's only a narrow band in space that is just right, and Earth's first lucky break was to be born right in the middle of it. Comfortably ensconced in this so-called habitable zone, we fortuitously found ourselves basking in the warmth of just the right kind of star. Thanks to its size, the Sun has emitted energy at a fairly steady rate for a very long time, giving life plenty of time to develop. Added to that was the happy accident of Jupiter's formation. The giant planet stands guard over us, its massive gravitational field drawing potentially deadly projectiles away from Earth-bound trajectories. If Jupiter had been smaller or more distant, Earth would have been wracked more often by colossal collisions, and life would probably have been snuffed out if it had even managed to get off the ground. Then there was the chance collision with Earth's twin planet, Theia, the wreckage of which became our travelling companion. The Moon's reassuring tug steadied Earth's erratic tilt, keeping climate within comfortable limits, and it set the familiar rhythms of seasons and tides.

Theia bequeathed another parting gift to our infant planet – its core. The extra mass made Earth big enough to hold on to a substantial atmosphere, in turn creating the greenhouse blanket that traps the Sun's warmth. And from the atmosphere, of course, came water. Mars and possibly even Venus may once have had liquid water at the surface, but Earth is unique in keeping its oceans for so long. As far as we know, no other body in the solar system has so much as a puddle, yet Earth has retained its enormous seas for more than 4 billion years. When we look out across them, we see a view that has remained unchanged almost since the birth of our planet, and that is truly remarkable. It is this capacity to maintain water in liquid form that has made our planet an ark for life.

Being just the right distance from the Sun did not, on its own, save the early Earth's water from boiling away into space or freezing solid. The planet also needed a climate-control system, but for that to work yet another unique feature was required: plate tectonics. Our neighbouring planets may have had moving crusts in the distant past, but for the last billion years or so, their surfaces have been stagnant and largely still. Volcanoes are pretty common in the solar system, but no other planets have the distinctive linear mountain chains that suture Earth's broken and shifting shell.

OPPOSITE **THE VERY LARGE ARRAY (VLA) RADIO TELESCOPES IN NEW MEXICO HAVE BEEN PART OF THE SCIENTIFIC SEARCH FOR EXTRATERRESTRIAL LIFE, THOUGH SO FAR NOTHING HAS BEEN FOUND.**

And the reason may simply be that those planets have lost their water. On the young Earth, sea water seeped into the crust, weakening the brittle rock and making it supple enough to bend and break. Heat rising from deep inside the planet split the weakened crust into fragments and set them moving, making them collide and buckle, turning Earth's skin into a conveyor belt of destruction and renewal. The moving plates provided an outlet for pent-up internal heat, and so the convection currents from the deep interior strengthened, thereby exporting more heat from the core. Earth's convective engine was now purring healthily, strong enough, in fact, to create the planet's magnetic field. That, in turn, would cast a protective shield around Earth, screening the surface from cosmic radiation and reducing the loss of atmospheric gases into space.

Water and moving plates conspired to get Earth's climate control going by creating landmasses that stuck up from the oceans. The first continents were born when wet oceanic crust was forced down into the mantle, where the water and molten rock combined to form a more buoyant type of rock: granite. As soon as the granite landmasses rose above the waves, they began to weather away, releasing minerals that gradually neutralized the acidic seas and made the waters more hospitable to life. And when weathering began, so did the inexorable drawdown of carbon dioxide from the atmosphere. Earth's global thermostat now began to operate, stabilizing the temperature at just the right level for water to remain liquid.

If things had been slightly different, none of it would have worked. If Earth had been smaller, there wouldn't be enough gravity for a thick atmosphere and a strong greenhouse effect, and the surface would be too cold for liquid water. Without water, the shifting plates would freeze solid too. If Earth were bigger, the atmosphere would be too dense and the climate baking hot. The amount of water had to be just right too. If there was too little, the world would be packed with continents, the weathering of which would draw so much carbon dioxide out of the atmosphere that the planet would freeze. Too much water and the continents would be submerged, making weathering impossible. Either way, without temperature control, Earth's water would either evaporate away into space or freeze.

It is incredible to think what a complicated and unlikely chain of events was needed to create a planet suitable for life. Earth had to be the right size and the right distance from the right kind of sun. It needed the right collision to create a moon for stability, the right protective planet in the wings, and just the right amount of water delivered from space. Add to that the right range of building materials, the correct amount of heat to make plate tectonics work, and the right chemistry to make our atmosphere and oceans conducive to life. All told, it is quite a feat to build a planet that is home to life – well, life that is more than simply bacterial slime. A planet like Earth might be very rare indeed.

But it is not simply this mix of cosmic fluke and planetary good fortune that has made our world so special. Earth has another trick up its sleeve, and that is the extraordinary way in which the planet and its inhabitants seem to have struck up a partnership.

Mother Earth

In the 1960s, the British scientist James Lovelock worked as a consultant on the Viking missions that NASA was planning to launch in the next decade to search for life on Mars. Well before the Viking landers touched down and began to analyse the dusty red Martian soil, Lovelock realized there was another way to detect life on an alien planet – one that didn't involve the expense of sending spacecraft. All you needed to do was find out the composition of its atmosphere, and that could be done by analysing the light captured with a telescope. On a lifeless planet, the atmosphere would be an inert mixture of unreactive gases left behind after all possible chemical reactions had ceased. In the atmospheres of Venus and Mars, for instance, everything that could react has reacted. The chemistry of a living world

ABOVE **THE INDEPENDENT BRITISH SCIENTIST JAMES LOVELOCK, WHOSE GAIA HYPOTHESIS PUT FORWARD THE CONTROVERSIAL IDEA THAT EARTH IS A LIVING ORGANISM.**

is different: organisms continually recycle the elements necessary for life as they take in certain chemicals from their environment and expel others, using the atmosphere as a conveyor belt. This gives the atmosphere a highly unlikely chemical signature – one that couldn't be maintained without life. On Earth, the air we breathe is an anomalous mixture of oxygen and nitrogen, with a trace of carbon dioxide, methane and other gases. The oxygen ought to react with the methane and nitrogen and disappear, yet these unstable bedfellows coexist happily. It is a marriage of convenience made possible by the constant recycling efforts of life.

Lovelock's idea that organisms regulate their environment would develop over the following decades into a remarkably audacious and all-encompassing vision of how our planet works. Central to this vision is the notion of 'homeostasis', a medical term that describes how a healthy body automatically keeps temperature, water balance and many other variables at close to optimum levels, regardless of changing conditions outside. Lovelock argued that our planetary environment is similarly homeostatic, with the temperature and atmospheric composition kept stable by life. The planet's organisms and the environment they create form a single, self-regulating entity that keeps our planet in a comfortable state, fit for the propagation of life. This provocative idea – that the largest organism on Earth is Earth itself – needed a name, so Lovelock gave it that of the Greek mother-Earth goddess: Gaia.

For almost 4 billion years, even though the Sun has grown hotter and terrible environmental crises have come and gone, the planet has almost never departed from the narrow range of conditions required to sustain life. The average temperature has stayed between 0°C (32°F) and 100°C (212°F); concentrations of phosphorus, nitrogen and sulphur have always been adequate; poisons have been kept in check; the sea has never been more salty than life can withstand; oxygen has been maintained at tolerable levels; and, most crucially of all, water has been prevented from leaving the planet. For Lovelock, it is Gaia – the intricate web of life and environment – that has been our planet's steward.

For many geologists, the idea of Earth as a self-regulating 'superorganism' goes too far, but most accept Lovelock's core message that life does not merely inhabit Earth but is fundamental to the way it operates. Biological processes interact with physical and chemical ones and so have an enormous impact on planetary conditions. For instance, life has ensured that plate tectonics – so critical in making Earth habitable – hasn't stopped on our planet. The marine organisms that draw carbon dioxide into their bodies and then die to lock that carbon away in limestone sea floors ensure that the atmosphere's insulating blanket never gets too stifling and that water – which oils the wheels of moving plates – remains. Without life as the middleman, the global carbon cycle would break down and the climate would go haywire, making Earth uninhabitable.

So for hundreds of millions, perhaps even billions, of years, life has fine-tuned the climate to its own advantage. But there's a problem. It seems that life might have got too good at controlling its environment, and in the long run the same processes that sustained life might start to destroy it. And it seems that the demise of life on Earth has already started.

The End of the World

We are currently living in a short era of man-made global warming. This might sound like the beginning of the end for Earth's finely balanced climate, but in fact our planet has long been in decline. Earth is old. It has been a living planet for perhaps 4 billion years and it is already past its peak, the decline already under way – a decline that will have far-reaching consequences for humanity. Even our best (or is that worst?) efforts at changing the climate can only delay what lies in store for us. For in the long run, it isn't warming that will cause us problems but cooling. As we discovered in the previous chapter, we are still in an ice age, albeit in a brief warm lull that has allowed our civilization to flourish. Man-made global warming might prolong this Indian summer for a few thousand years, but the ice is sure to return soon after, and the end of the artificially warm era will be brutal.

The return of the glaciers will not threaten life itself – after all, life has survived many past glaciations. But if civilization still exists it will face enormous challenges feeding itself, because the world's agricultural productivity will plummet. The population during the last ice age was perhaps 2–3 million; by the next one it could be 100 billion. Humanity might well survive the inevitable starvation and warfare, but in vastly reduced numbers. And if we're around for long enough, we'll witness the ice retreat once more, only to come and go another 50 times or so.

Between 2 million and 10 million years from now, tectonic motion will free Earth from the grip of the glaciers, and the Age of Ice will finally give way to a period of regeneration. The continents of the northern hemisphere will drift south, reducing the amount of land available for glaciation, and Antarctica will move north, its melting ice raising sea levels by about 100 metres (330 feet). The world will warm up, and life will thrive – but only for a while. For on the horizon, storm clouds will be gathering and, remarkably, we have already begun to feel the first spots of rain.

The problem is something we're familiar with today: carbon dioxide. But there's a twist. The issue for the future world is that there will be too little of it, not too much. Levels of this greenhouse gas have been on the decline for several hundred million years, as geological processes have got ever more effective in locking carbon away in the rocks. The result has been a steady

ABOVE **THE GLOBAL SPREAD OF GRASSLANDS WAS EARTH'S LAST GASP ATTEMPT TO COUNTER FALLING CARBON DIOXIDE LEVELS, BUT CAN IT SAVE THE PLANET?**

drift towards cooler conditions that climaxed 2 million years ago at the peak of the ice age. Although future plate movements will free us from the ice, they can do little about the dwindling carbon dioxide.

Life itself is partly to blame. Around 600 million years ago, when complex life was beginning to get off the ground, carbon dioxide levels were far higher than today and on the rise. They peaked about 400 million years ago, when Earth's air contained 20 times as much carbon dioxide as it does now. Long before humans were around to meddle with things, the planet was confronting serious global warming. Faced with this challenge, life developed an amazing coping mechanism. Land plants evolved. As they spread across the continents, evolving from sparse, mossy forms to lush forests of towering trees, they sucked more and more carbon dioxide out of the atmosphere and stored it in the soil. Great quantities of carbon were locked up in rotting vegetation, later to become coal (from which the carbon would be liberated to fuel modern global warming). The forests also caused the land to weather more quickly, freeing minerals from the rock and thereby stimulating marine plankton, which drew even more carbon dioxide from the air and trapped it in limestone. Eventually, the success of plants had created the opposite problem: carbon dioxide levels were falling and the climate was cooling down. So the planet came up with a stopgap solution – it invented grass.

The spread of grasslands across the globe 40 million years ago represented the last desperate throw of

the dice in the face of carbon dioxide depletion. Grasses thrived on the increasingly meagre diet of carbon dioxide, and they expanded into new pastures as the planet cooled and dried with the onset of the ice age. But in time, perhaps a few hundred million years from now, carbon dioxide levels will drop below a critical threshold and even the sturdy grasses will struggle to survive. Plants will disappear, and Earth will become a desolate wasteland of rock and gravel, with huge, braided rivers – their banks no longer bound by roots – that strip the land of soil and wash it out to sea. Once the land is barren, few nutrients will flow into the oceans and they will starve too. With the loss of marine plankton and land plants, the lungs of the Earth will have withered away, and oxygen levels will nosedive. This is the final tipping point – the moment at which life loses control of Earth's bodily functions, and the planet's life-support systems begin to shut down. A few tens of millions of years after the downfall of plants – essential for the production of oxygen – less than 1 per cent of the atmosphere will be oxygen, compared to 21 per cent today. Any animals still around would suffocate, and the ozone layer would vanish, allowing high-frequency UV rays to sterilize the planet's surface.

As an abode for life, Earth is at best in its late middle age and more probably in its old age. Our current 'Age of Animals' is a last hurrah, the highest point in the trajectory of life's history and the apogee of our planet's ecological complexity. Its end will mark the beginning of a relentless simplification of life on Earth. As time progresses, complex animal and plant fossils will be overlain by a succession of ever-simpler organisms, the history of a world dying, system by system. There will be ecological chaos amidst the dizzying evolutionary procession of forms scurrying about the *Titanic*'s deck, all trying – to no avail – to evolve life

LEFT **THE PARCHED EARTH OF DEATH VALLEY IN CALIFORNIA – A GLIMPSE OF AN EXTREME, INHOSPITABLE WORLD THAT WILL RETURN IN THE DISTANT FUTURE.**

jackets. Life will retreat to the deep sea, but even this move will not be enough to save Earth's residents from their inevitable fate.

Around a billion years from now, with the Sun 10 per cent brighter than today, the average global temperature will be near 70°C (158°F). The seas will evaporate into the atmosphere, leaving wildly coloured pools of putrid brine scattered across vast plains of salt. The only organisms left will be halophiles – microbes that live inside tiny water inclusions in salt. Bacteria will once more have the world to themselves, bringing evolution full circle.

The loss of the oceans will be the single most important step in the annihilation of life, causing a runaway greenhouse effect and soaring global temperatures. 'Extremophile' bacteria may be remarkable in their ability to flourish in scalding volcanic springs, but they are not remarkable enough to endure the hellish temperatures that will come. Life's basic chemical processes cannot function above 112°C (234°F), but after the oceans disappear Earth will become a clone of Venus, where the surface temperature of 450°C (842°F) is hot enough to melt lead.

Without water, Earth's crust will lose its flexibility and tectonic plates will grind to a halt. Like the stopping of a human heart, the failure of plate tectonics will have everlasting effects. Linear mountain chains will stop forming, and the ocean basins will fill with sediment eroded from the continents. The planet will become flatter. The crust will thicken, like that of Venus or Mars, and heat will build up underneath. Then, as on Venus, the whole of the Earth's crust will occasionally melt, turning Earth into a new hell and obliterating every trace of life's existence.

Already a roasting desert, Earth will continue to heat up as the ageing Sun burns ever hotter. Over the past 4 billion years, the Sun's output has increased by 30 per cent, and the trend will carry on until Earth is consumed. About 5 billion years from now, as the Sun's hydrogen fuel becomes depleted, it will swell into a red giant. To see this from Earth would be to see the Sun completely filling the daytime sky. Its radiation will be 2000 times more powerful than now, causing Earth's surface to heat up beyond 2000°C (3630°F) – hot enough to melt mountains and smooth over the entire planet.

It is truly a cataclysmic end for our planet. As the Sun flares out, its atmosphere will produce a drag force on the Moon, forcing it to spiral inwards and collide with Earth, the planet that gave birth to it so many billion years before. The expanding Sun will become as much as 6000 times brighter than today, its diameter pulsing out to the size of Earth's orbit. Earth will cease to exist.

Thankfully, this won't happen for 7 billion years or so. Meanwhile, we humans have some more pressing concerns.

The Sixth Extinction

If you ever doubted the scale of humanity's impact on the planet, just look at Earth from space. At night, electric lights mark out our conquest of the globe with millions of pinpoints, blazing swathes of them covering Europe, Japan and eastern North America. In fact, our reach extends even further, so that no corner of the planet – not even the snowy wastes of the polar extremities – is exempt from our influence. Never has a species dominated Earth so completely.

It is ironic that just as we begin to understand how fundamental life is to the workings of our planet, we are systematically wiping out the very biodiversity that makes Earth such a special, possibly unique, world. Some 30–50 per cent of Earth's land surface has been transformed and degraded by human exploitation, causing a staggering loss of natural habitat. According to the World Conservation Union, nearly 16,000 species are now on the brink of extinction, including one in four mammal species, one in three amphibians and one in eight birds. In the last 500 years, human activity has caused more than 800 confirmed extinctions, but the true situation is probably far worse than this figure implies. Only around 1.75 million of the

ABOVE ONCE A VAST CARPET OF HEALTHY VEGETATION (RED), THE AMAZON RAINFOREST IS NOW DISFIGURED BY AN UGLY PATCHWORK OF DEFORESTATION. HERE IN BOLIVIA, LOGGERS HAVE CUT LONG PATHS INTO THE FOREST, RANCHERS HAVE CLEARED LARGE BLOCKS FOR THEIR HERDS, AND VILLAGES HAVE RADIAL ARRANGEMENTS OF FIELDS AND FARMS.

estimated 13–14 million species on Earth have been described, and we may well be losing undiscovered species at a rate of hundreds a week. As long ago as 1993, Harvard biologist E. O. Wilson estimated that Earth loses around 30,000 species per year, which works out as three species per hour. At that rate, half the world's species will be gone by the middle of the century.

Biologists are calling it the Sixth Extinction. As we've seen earlier in the book, there have been at least five occasions in Earth's history when some kind of catastrophe has pushed life on Earth to the brink of collapse, wiping out an enormous proportion of the planet's species in one fell swoop. The fossil record is also littered with many less severe calamities in which significant numbers of animals and plants went to the wall together. Of course, species come and go all the time as a natural part of evolution's turnover, and most of the species that have ever lived are now extinct. From fossil records painstakingly gleaned from Earth's rocks, geologists estimate the background rate of extinction to be about one species every four years, which is surprisingly frequent. Even mass extinctions could be viewed as being part of the natural order. Some geologists argue that the occasional humungous outpouring of lava, asteroid collision or poisoning of the ocean presses the evolutionary reset button, clearing out life's

closet and allowing new ecological opportunists to arise phoenix-like from the cosmic or volcanic ashes.

So, on the face of it, colossal culls would seem to be part and parcel of a thriving ark of planetary life. But what concerns many biologists is that the biodiversity loss that humans are inflicting on the planet today may be happening faster than ever before. If 30,000 species are indeed disappearing every year, then the current extinction rate is 120,000 times higher than the natural background level. More conservative estimates suggest the loss of biodiversity is happening only 100–1000 times faster than the prehuman rate. More significantly, some geologists reckon that species are vanishing faster than during the 'big five' mass extinctions. Humans, it seems, are well on the way to becoming a destructive force to match asteroid strikes and supervolcano eruptions. And we are putting unprecedented demands on our planet's coping mechanisms.

Forests are on the front line of humanity's battle with nature. As the world's woodlands are felled to make way for agriculture and living space, the loss of natural habitat is leading to more and more extinctions. Of course, humans have been clearing forests for millennia, but it is only in the recent era of exponentially growing population that habitat loss has emerged as the prime cause of extinction. And disappearing species is not the only problem that deforestation brings. Forests are living storehouses of carbon and play a critical role in removing carbon dioxide from the atmosphere. The full implications of their wholesale destruction remain unclear, but one thing is sure: we are damaging a vital part of our planet's atmospheric thermostat – one of its most finely tuned life-support systems. And what's happening on land is starting to happen in the sea.

OPPOSITE **THE ICONIC BAOBAB TREES OF MADAGASCAR – THE ISLAND'S NATIONAL SYMBOL – ARE STILL STANDING, BUT THE TROPICAL FORESTS THAT ONCE CLOAKED MUCH OF THE ISLAND HAVE MOSTLY DISAPPEARED.**

Our seas have long been spared the worst excesses of human exploitation, but this is starting to change. In the last century, mechanized fishing increased the world's annual harvest of fish from 5 million to 90 million tonnes. Our species now dominates the marine food chain, consuming more than 25 per cent of the marine ecosystem's primary production in upwelling regions of the ocean and 35 per cent in temperate continental shelves. In many regions, species that once flourished in great numbers are now rare or gone completely. For thousands of years, we have used the oceans as a dumping ground, but their ability to mop up our rubbish and pollution is now beginning to look limited. Sea water has soaked up an estimated one-third of all the carbon dioxide humans have emitted since the beginning of the Industrial Revolution (some suggest it could be half), but natural processes are struggling to keep up with the rate at which we are pumping it out. The result is that sea water is becoming increasingly acidic. Another of the planet's key self-regulatory mechanisms is under attack.

All things considered, it looks pretty bleak for our planet. Earth may have been born lucky and may have evolved a marriage with life to keep itself going, but we weren't part of the original design. This is the first time that any species has knowingly altered the planet – and interfered with so many different Earth systems at once. The oceans and forests might do all they can to remove the carbon dioxide we release, but our actions are simply overwhelming them. We're pumping out pollution too fast for them to keep up. Who knows what will happen when we mess with our delicately balanced atmosphere, but it seems that humans have the potential to dismantle the fundamental ties that bind our planet together. Have we become a lethal threat to the beautiful planet that nurtured us? Perhaps not. You only have to go back to the site of the last great extinction event – the one that extinguished the dinosaurs and gave our mammalian ancestors the opportunity to take over – to get a rather different perspective on the dangers facing planet Earth.

Near-death Experiences

Everywhere you go in the low, scrubby jungle of Mexico's Yucatán peninsula, you find strange holes in the ground. They're called *cenotes*, and there are literally hundreds of them. Most have never been explored. Some are filled to the brim with water – circular lakes that pockmark the verdant forest – but others are simply narrow, gaping entrances to a deeper water table and a remarkable flooded subterranean world. For here, in the relatively young limestone rocks that underpin the entire Yucatán region, is one of the planet's most extensive cave systems. As divers have explored this maze of drowned caverns, they have gradually revealed a massive network of interconnected tunnels and passageways that runs for hundreds of kilometres. Most remarkable are the *cenotes* in the northwest corner of the peninsula, which link up in a chain of especially deep chasms, some hundreds of metres deep – essentially bottomless to the divers that venture into them. But the real story of these caves can't be gleaned in the pitch-black, watery abyss. Instead, you have to look down on the Yucatán from high above. Images from space reveal that the deep *cenotes* form an arc around the ancient, buried rim of the Chicxulub impact crater. Deep below, great fissures and fractures ripped open in the blast served as persistent pathways for groundwater, which slowly etched out the caves and passageways. In other words, Yucatán's remarkable ring of *cenotes* is scar tissue that formed as the open wound of the giant impact gradually healed and became buried.

Sixty-five million years on, there is nothing obvious to suggest that Yucatán was the site of one of the most momentous events in the history of life. The mysterious *cenote* ring offers a faint clue, but all other visible signs of the enormous impact have long been erased: first by shallow seas that buried the crater beneath muddy sediment, and later by the soils and vegetation that gradually carpeted the emerging coastal plain. Today, hundreds of kilometres of unbroken jungle covers the flat Yucatán lowlands. Scattered within it are strange mounds of stones that testify to another lost world: the Maya civilization. The ruins of their cities now lie hidden under the scrubby forest, but their ceremonial stone pyramids still stand, the legacy of a huge empire that stretched north into Mexico and south as far as El Salvador. The Maya peoples ruled this vast heartland for almost 1000

ABOVE **THE MAGNIFICENT SUBTERRANEAN WORLD OF THE CENOTES OF MEXICO'S YUCATÁN PENINSULA – A LEGACY OF ONE OF LIFE'S MOST TRAUMATIC NEAR-DEATH EXPERIENCES.**

years, but by the late eighth century their world was faltering. Severe droughts that lasted years or decades crippled their agriculture, and in city after city – with authorities powerless to bring the rains – civil unrest erupted. By AD 900, lowlands that had bustled with 8–10 million people a century earlier were virtually stripped of their population by famine and disease. Ironically, the virtually waterless northern lowlands of Yucatán were the last area to succumb to the deteriorating climate, fed as they were by their sunken freshwater pools and wells. But even the *cenotes* could not halt the Mayans' decline. Those who survived were dispersed into small hamlets, and little by little, the jungle reclaimed its former territory. Today, the occasional Mayan monuments that rise out of clearings in the Yucatán forest seem to offer a double message: no civilization will last for ever, and no civilization will outlast nature.

It is difficult to visit Yucatán and not come away amazed by the sheer resilience of life on planet Earth. This is where a great civilization prospered for almost a millennium, yet there is virtually no sign of it. This is where a gigantic meteorite exploded, but even that event has left hardly a trace. Earth – and life – recovered. It is this ability to deal with catastrophe that is truly special about Earth. Our planet is really tough. It has been robust for billions of years, and there is nothing to suggest that this is going to change any time soon. In the long run, Earth can probably cope with anything we throw at it. We could clear the jungles, but as the Mayan ruins attest, a jungle can return in a matter of centuries. We can poison the oceans, but they will undoubtedly recover, just as they have done after being starved of oxygen and asphyxiated in the past. We could burn all Earth's fossil fuels and flood the atmosphere with carbon dioxide, but even then it would take the planet only a few million years to lock the carbon away again. Even the creatures we're systematically wiping out will be replaced as evolution works its magic. Extinction is death, but it is not the end of birth. Earth and life will bounce back. It's just a question of time, and the planet has plenty of that.

But that's not to say that all the rapid changes we're forcing on Earth don't matter. Humans operate on a different timescale to the planet. We evolved to occupy the world as it is – an Earth with coral reefs, rainforests, and ice at the poles. In changing this world, we are altering the very environment that allowed our species and civilization to flourish. Just like the dinosaurs that exited stage left as the meteorite landed at Chixculub, we can't cope with sudden change. And this time it is us, not the dinosaurs, who are top of the food chain.

So, all this stuff about 'saving planet Earth'. That's nonsense. That's not the problem. Planet Earth doesn't need saving. For four-and-a-half billion years, Earth has been a survivor. It's not the planet we should be worrying about. It's us.

RIGHT **MEXICO'S YUCATÁN JUNGLE IS GRADUALLY ENCROACHING ON ANCIENT MAYAN MONUMENTS – RELICS OF A LONG LOST WORLD.**

Timeline

Since its earliest years, Earth has switched back and forth between two kinds of world: one that is completely ice-free and enjoys a balmy greenhouse climate, and one in which ice covers a significant proportion of the planet. The fluctuation between greenhouse and ice-house states is driven by the interplay of several different processes, including rhythmic wobbles in Earth's tilt and orbit, and the ever-changing arrangement of continents and oceans. All that we know of these past climatic oscillations comes from ancient rocks and from the fossilized remains of life forms preserved within them. It is this rock record that has allowed the major geological stages (periods) in Earth's last few hundred million years to be established. A broad picture of Earth's past climate is known for the last 550 million years or so, but further back – prior to the emergence of complex life at the start of the Cambrian Period – the picture is extremely hazy, though there is good reason to think that Earth was in an icehouse state throughout much of the Precambrian. Seen in this long-term perspective, the present global warming may seem inconsequential, but remember that we humans have never experienced an ice-free world. Note that the timeline is not drawn to scale – the Precambrian makes up nearly 90 per cent of Earth's history but is allocated only a 15 per cent share of the timeline.

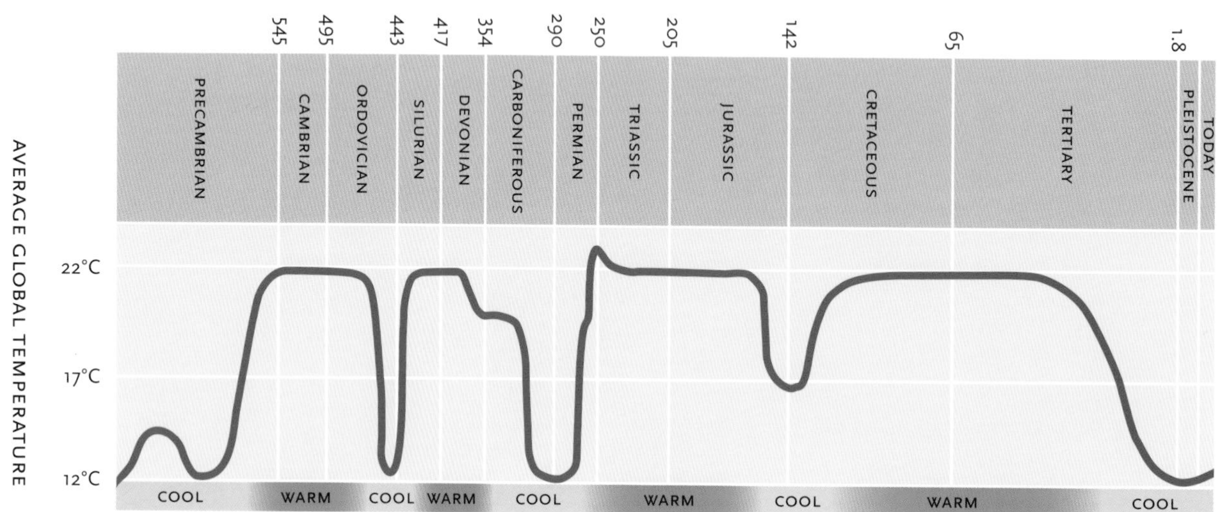

Further Reading

Ball, P. *H_2O: A Biography of Water* (Weidenfeld & Nicolson, 1999)
An examination of the baffling and varied nature of water from a leading popular science writer.

Beerling, D. *The Emerald Planet: How Plants Changed Earth History* (OUP, 2007)
A fascinating tale of 'green evolution'.

Bryson, B. *A Short History of Nearly Everything* (Doubleday, 2003)
A delightful, irreverent and constantly revealing romp through all that we don't know about the Cosmos, our planet and the life on it.

Catermole, P. *Building Planet Earth: Five Billion Years of Earth History* (CUP, 2000)
A readable, succinct and well-illustrated overview of how our planet came to be and how it works.

Fortey, R. *The Earth: An Intimate History* (HarperCollins, 2004)
The latest tales of the planet from one of the best communicators of geology around.

Hancock, P. and Skinner, B. (eds) *The Oxford Companion to the Earth* (OUP, 2000)
A comprehensive but technical coverage of most of the dominant topics and themes of modern geological thought.

Lane, N. *Oxygen: The Molecule that Made the World* (OUP, 2002)
The remarkable relationship between life on earth and the element that is critical for its survival.

Lynch, J. *Wild Weather* (BBC Books, 2002)
A popular look at the forces of heat, cold, wind and rain that make up the world's weather system.

Margulis, L. *The Symbiotic Planet* (Weidenfeld & Nicolson, 1998)
A controversial but very readable account of how Earth functions and how this relates to the Gaia concept of a living Earth.

Stewart, I. *Journeys from the Centre of the Earth* (Century, 2005)
A look at how geology and culture have combined to shape the history of the Mediterranean.

Taylor, S. R. *Solar System Evolution* (CUP, 2001)
A rather dense textbook but an authoritative summary of the latest understanding of how Earth and its planetary neighbours came to be by one of the leading researchers in the field.

Van Andel, T. *New Views on an Old Planet* (CUP, 1994)
An easy-to-follow, informal outline of the major changes in Earth history by a leading geologist.

Ward, P. D. and Brownlee, D. *The Life and Death of Planet Earth* (Piatkus Books, 2003)
A very readable account of how the planet may end by a leading geologist and astronomer team.

Ward, P. D. and Brownlee, D. *Rare Earth: Why Complex Life is Uncommon in the Universe* (Copernicus Books, 2000)
A contentious but well-argued explanation of how advanced life on Earth depends on special circumstances predicated by its restless, and occasionally downright violent, geological past.

Westbroek, P. *Life as a Geological Force* (Norton & Co.,1991)
One of the first books to popularize how geological systems interact to keep Earth habitable.

Wilson, R. C. L., Drury, S. A. and Chapman, J. L. *The Great Ice Age: Climate Change and Life* (Routledge/Open University, 2000)
An authoritative work on the climate changes of the last 2.5 million years.

Index

Page references in *italic* refer to illustrations

Acknowledgements

The television series on which this book is based was conceived by John and presented by Iain, but it was shaped and created by an enthusiastic production team led by Philip Dolling and Jonathan Renouf. The research team of Oliver Page, Rebecca Harrison, Ben Wilson, Roeland Doust and Liz Vancura turned stones into stories, which were brought to life by directors Annabel Gillings, Matt Gyves, Sophie Harris, Ben Lawrie and Paul Olding, and the sprawling television project was managed by Paul Appleton with Rachel Dulin, Anya Roberts, Shelley Raichura, Alison Seymour and Anja Zoll-Khan. Special thanks are due to Jonathan, Paul, Ben and Rebecca for reading and commenting on early sections of the book.

Iain would especially like to thank the film crews whose support enabled him to write so much of this book on location, especially Tim Cragg and Adam Prescod, who bore the brunt of the extreme travels. Also, his colleagues in Geology at Plymouth University provided amazing cover and encouragement during this year of geological globetrotting, with special thanks due to Mark Anderson, Jim Griffiths, David Huntley, Roddy Williamson and Mark Cleary.

Countless scientists and specialists gave time and effort to the preparation of the series and book, but some who deserve special thanks for help and guidance above and beyond the call of duty are Phil Bland, Doug Benn, Jim Edds, Brian Fisher, Clive Oppenheimer, David Price, Winston Seiler, Richard Twitchett and Tim Wright.

For the book, we are indebted to the commissioning efforts of Martin Redfern, the project supervision skills of Christopher Tinker and the design talents of Andrew Barron. Special credit goes to Ben Morgan for ironing out the rough edges in our draft chapters and saving us the embarrassment of the odd error here and there.

Finally, it is simply not possible to undertake a challenging project like this without the continued support of friends and family. Iain's peripatetic existence over the year has been tolerated magnificently by Paola and the kids, Cara and Lauren. And John, as ever, has been humoured by Ewa, Christopher and Toby. We hope they all feel it was worth it!

Picture Credits

BBC Books would like to thank the following for providing photographs and for permission to use copyright material. While every effort has been made to trace and acknowledge copyright holders, we would like to apologize should there have been any errors or omissions.

ALAMY: Aerial Archives page 191; Jon Arnold Images 110; Blickwinkel 64–5; Tim Graham 78; North Wind Picture Archives 50 (left); Pacific Press Service 86; Mark Pearson 148–9; Photo Resource Hawaii 67 (top); The Print Collector 15, 50 (right); Profimedia Internation s.r.o. 96; Kevin Schafer 1, 89; Stock Connection 84–5; Woodfall Wild images 111; WoodyStock 99
ART DIRECTORS/TRIP: Mary Jelliffe 166–7, 199 (background)
BBC: 60, 71, 108, 140, 182–3 (background), 190 (background), 199 (background), 212 (background); Tim Cragg 147; Matthew Guyves 76, 175; Rebecca Harrison 188–9, 196–7; Joe Jennings 106–7 (all); Ben Lawrie 91 (inset), 226; Paul Olding 72–3, 79; Frederick Olivier 44 (bottom); Iain Stewart 28, 66 (top, repeated as backgrounds on 68, 77, 87, 91 and 96), 67 (bottom), 80–1, 94, 97, 136–7, 146, 212, 217; Marek Zuk 230–1
BBC BOOKS: 74, 158, 232
WILSON BENTLEY: 183
GARY BRAASCH: 209
CORNELIUS BRAUN: 190
COURTESY OF THE ARCHIVES, CALIFORNIA INSTITUTE OF TECHNOLOGY: 51 (bottom left)
CORBIS: Martin Alipaz/epa 164–5; Tom Bean 186–7; Gary Bell/zefa 170; Jonathan Blair 24, 169; China Newsphoto/Reuters 152–3; CNP 133; Alejandro Ernesto/epa 132; Michael Freeman 20–1; Alberto Garcia 100–1; Aaron Horowitz 36–7; Peter Johnson 221; Karen Kasmauski 26–7; Frans Lanting 16–17; Frank Lukasseck 222; J. Marshall/Tribaleye Images 105; Arthur Morris 112; Louise Murray/RHPL 122–3; John Nakata 150–1; Kazuyoshi Nomachi 134; Charles O'Rear 54–5; Douglas Peebles 2–3; Jim Reed 126–7; Reuters 162 (bottom); Paul Souders 178–9; Jim Sugar 83; George W. Wright 219; Jim Zuckerman 13
ECOSCENE: Edward Bent 66 (bottom)
FLPA: Matthias Breiter/Minden Pictures 69; Frans Lanting 4 (centre), 62
GALAXY: 4 (right), 35, 40, 49, 52, 102, 115
GEOLOGICAL SOCIETY: 51 (right)
GEOPHOTOS: 181, 184–5
GETTY IMAGES: Macduff Everton 228–9
NASA: 4–5 (background), 5 (right), 6, 8, 11 (repeated as backgrounds on 15, 18 and 19), 14, 18–19, 23, 33, 34, 38, 42–3, 56, 58–9, 68, 71, 82, 87, 93, 109, 125, 130, 131, 142, 145, 156, 173, 174, 192–3, 194, 202–3, 205, 207, 213, 214, 225, endpaper (back)

NATIONAL MUSEUM OF THE USAF: 104
PHOTOLIBRARY: 206; Karen Jettmar 5 (centre), 176; Ken Sherman 182
PHOTOSHOT: 143
PLANETARY VISIONS: 141
PETER B. RHINES: University of Washington 160–1
SEAWIFS/NASA: 199
H. SLUPETSKY/UNIVERSITY OF SALZBURG: 208
SPL: 156 (background), 162 (background); Michael Abbey 44 (top); Doug Allan 200–1; John Chumack 4 (left), 30; Ted Clutter 121 (top); Christian Darkin 75; G. Brad Lewis 5 (left), 138 (repeated as backgrounds on 146, 158, 162 and 172); M-Sat Ltd 58, 135; NASA 38 (background), 44 (background), 50 (background), 51 (background), 118; Alfred Pasieka 172; Alex Rosenfeld 77; Detlev van Ravenswaay 45; Jeremy Walker 154–5; Zephyr 114
STILL PICTURES: 210; BIOS/John Cancalosi 116–17, 120; J. Mallwitz/ Wildlife 121 (bottom); Jim Wark 195
GILBERT WALKER INSTITUTE: 162 (top).
RUSSELL WHITE: 128–9 (repeated as backgrounds on 105, 112, 114–5, 118, 123, 132 and 134)
WHOLEWORLDFORDESERTSANDMAPS: 123, endpaper (front).

PAGE 1 The northern lights (aurora borealis) seen from Canada.
PAGES 2–3 Mount Kilauea erupting.
ENDPAPERS Composite satellite images of Earth by day (front) and by night (back).

This book is published to accompany the BBC television series *Earth: The Power of the Planet*, first broadcast on BBC2 in 2007.

Executive producer Philip Dolling
Series producer Jonathan Renouf

Published in 2007 by BBC Books, an imprint of Ebury Publishing.
A Random House Group Company

10 9 8 7 6 5 4 3 2 1

The Random House Group Limited
Reg. No. 954009

Addresses for companies within the Random House Group can be found at www.randomhouse.co.uk

A CIP catalogue record for this book is available from the British Library.

ISBN 978 0 563 53914 8

The Random House Group Limited supports The Forest Stewardship Council (FSC), the leading international forest certification organization. All our titles that are printed on Greenpeace approved FSC certified paper carry the FSC logo. Our paper procurement policy can be found at www.rbooks.co.uk/environment

Commissioning editor Martin Redfern
Project editor Christopher Tinker
Copy-editor Ben Morgan
Designer Andrew Barron
Picture researcher Joanne Forrest Smith
Illustrator (pages 74 & 158) Jerry Fowler
Production David Brimble

Colour origination and printing by Butler & Tanner, Frome, England.